THE MANUFACTURE OF MEDICAL AND HEALTH PRODUCTS BY TRANSGENIC PLANTS

THE MANUFACTURE OF MEDICAL AND HEALTH PRODUCTS BY TRANSGENIC PLANTS

Esra Galun
The Weizmann Institute of Science, Israel

Eithan Galun
Hadassah University Hospital, Israel

ICP Imperial College Press

Published by

Imperial College Press
57 Shelton Street
Covent Garden
London WC2H 9HE

Distributed by

World Scientific Publishing Co. Pte. Ltd.
P O Box 128, Farrer Road, Singapore 912805
USA office: Suite 1B, 1060 Main Street, River Edge, NJ 07661
UK office: 57 Shelton Street, Covent Garden, London WC2H 9HE

British Library Cataloguing-in-Publication Data
A catalogue record for this book is available from the British Library.

THE MANUFACTURE OF MEDICAL AND HEALTH PRODUCTS BY TRANSGENIC PLANTS

ISBN 1-86094-249-0
ISBN 1-86094-254-7 (pbk)

Printed in Singapore.

This book is dedicated to the late

Erna-Ester and David Mendel Galun,

our parents and grandparents

and to the late Arijeh-Baruch Galun,

our brother and uncle

Preface

קנאת סופרים תרבה חכמה!

The Envy of Scholars Will Increase Wisdom!

(Talmud, Bavli, Baba Batra, Page XX1/1)

The verbatim translation of the motto, that is an ancient Hebrew maxim, is stated above. But this maxim was and is used in a much wider context. Its broader meaning is that competition between creative minds will result in human progress. This progress encompasses a wide range of human endeavor such as literature, visual arts, music, scientific knowledge and, in our case, also biotechnology. At first sight, it looks strange that a competition, based on the exploitation of strong human instincts, such as jealousy and greediness and the strive for spiritual immortality, can result in human progress. But those who phrased the laws of intellectual property rights (using such concepts as novelty, non-obviousness and benefit, as conditions for such rights) knew their trade and understood human nature; the laws of these property rights were instrumental in recruiting human minds for the benefit of humanity. The competition between investigators ("scholars") contributed to the impressive progress in the field of the production of therapeutics by transgenic plants.

In the Preface of our preceding book, "*Transgenic Plants*" (Galun and Breiman 1997), we recalled the words of William Faulkner: "... to create, out of the human spirit something that did not exist before..." and we related this saying to the production of transgenic plants. We thus expressed the notion, with a hidden smile, that "... a plant biologist engaged in the production of transgenic plants may experience the feeling of being a creator". In the present Preface, we can use the Hebrew expression קל וחמר (Kal Va'chomer), which means an obvious and logical inference from minor to major. In our context, we can make the analogy that if producing a transgenic plant that did not exist in nature is giving its producer a feeling of being a creator, then if such a transgenic plant integrates and expresses a gene that is by itself a novel creation by the investigator, the feeling of being a creator is sharper. We are currently witnessing an ongoing interest in research work that aims to construct transgenic plants that have the potential to lead to the manufacture of medically beneficial products. Is this because such products are expected to be of commercial value or is it the prospect of becoming "creators" that is the drive behind these research efforts? Possibly both, and together with progress in biotechnology, they caused the interest in the subject of this book.

In the preceding book, our focus was on plants: how to improve crop production by genetic transformation that may lead to increased yield and improving its nutritional quality. In this book, we shall put man in the forefront. Obviously, the term "man" is defined by us, as in the Webster's dictionaries, to mean human being and it includes both females and males. Thus, this book shall attempt to describe how genetic transformation of plants may contribute to the health of man and animals. This contribution can be by the guarding of man against pathogens and by the availability of specific products that have therapeutic value in medical treatments against metabolic and physiological deficiencies that are either inherited or acquired maladies. Thus, after a brief discussion of plant molecular genetics and an introduction to immunology, we detail how specific human (and veterinary) antibodies can be manufactured by transgenic plants and also how vaccines against specific pathogens can be produced by genetic transformation of plants. Another part of this book handles the manufacture of various therapeutic compounds by transgenic plants. When plant was our focus, the safety issue was mainly on how to avoid damage to the plant environment by gene-flow from transgenic plants. When our emphasis is on human health, we have additional public acceptance and safety considerations.

While striving to keep to accuracy and providing a full account on the subject of this book (but avoiding technical protocols), we aimed at a wide range of readers. Therefore, we explained genetic, immunological, biotechnological as well as molecular genetic concepts and terms before using them. The book should therefore be understood by anyone with a basic biological knowledge. We believe that people interested in the state of the art of manufacturing therapeutic products by transgenic plants will benefit from this book. The book should also serve students of biotechnology and especially plant biotechnology, as an overview and reference book. Moreover, people who intend to explore this interesting field will gain, through this book, a knowledge of what was achieved and what the potentials of transgenic plants are as sources for products that can serve human and animal health-care.

Many individuals helped us during the writing and production of this book. Several members of the staff of the Department of Plant Sciences (The Weizmann Institute of Science, Rehovot, Israel) read parts of this book and provided valuable remarks: Professor M. Edelman, Professor H. Fromm, Professor G. Galili and Dr. G. Grafi. Professor A. Frensdorff (Tel Aviv University) and Professor D. Eilat (Hadassah University Hospital, Jerusalem) read the parts concerning antibody. Professor M. Galun (Tel Aviv University, wife and mother of the authors) made valuable remarks on several parts of the book. We also acknowledge the help, on specific subjects, provided by Professor E.C. Cocking, FRS (Nottingham University) and Professor R. Hagemann (University of Halle). Dr. N. Weinstein (Tel Aviv University) had important inputs on non-scientific issues mentioned in the book. We are grateful to staff members of the Imperial College Press, London, and the World Scientific Publishing Company, Singapore, for their efforts and contributions, and specifically to Mr. Alan Pui for his devotion. Our special gratitude is given to Mrs. Renee Grunebaum for her devoted typing and retyping of the manuscript. We are also grateful to the Departments of Photography and Graphic Arts of the Weizmann Institute of Science for their help with several of the figures in this book.

Let us recall the spirit of Aristotle (son of the great physician Nichomachus) who claimed that one of the two highest "goods" is to exercise intellectual life, that encompasses the study of Nature; and the spirit of Francis Bacon, who put forward the concept of stepwise progress in scientific knowledge. Thus, stepwise progress in the study of nature, and specifically in biotechnology and medical science, should lead us to overcome maladies and to improve the quality of life. We hope that this book will serve as an observation post for this stepwise progress.

E. Galun and E. Galun

Rehovot and Jerusalem

Contents

Chapter 1

Fundamentals of Plant Molecular Genetics

The last word of the title of this book, *plants*, requires a definition or rather a clarification of its meaning in our deliberations. By *plants*, we mean phanerogamous (i.e. flowering) plants. We may also mention other plants such as algae, mosses and ferns (cryptogamous plants). But then, we shall clearly indicate the latter type of plants, if they are mentioned. We shall commonly mention intact plants rather than specific organs (e.g. root culture) or cultured plant cells. There shall be one exception, in Chap. 5, where we shall deal with the culture of hairy roots.

The meaning of the term *transgenic plants* shall thus be plants that harbor alien genes. Again, the term *alien genes* requires an explanation or rather what shall be meant by it in this book. For us, *alien* will mean that it was introduced experimentally into the transgenic plant by genetic transformation rather than by other means such as introgression (cross-pollination whether intentionally or by nature) and mutagenesis.

While transgenic plants, by this definition, contain an alien gene (or several of them), it does not mean that the alien gene is expressed, leading to the production of either or both the respective messenger RNA (mRNA) and polypeptides. The alien gene may be a structural gene or a DNA sequence that carries a regulatory element, as shall be explained below. The transgene may be integrated into one of the plant's genomes; this is termed *stable transformation*. Or, it may not be integrated. The latter situation may cause *transient expression* (of mRNA and polypeptides). We should recall that plants contain three genomes. The major one is the nuclear genome where the genetic information is stored in the DNA of the chromosomes. Much less genetic information is stored in the chloroplasts' genome (*plastome*) and even less information is encompassed in the genome of the mitochondrium (chondriome). These last two genomes should not be ignored, albeit being poor in genetic information; they are vital for the functionality of plants. Furthermore, many of the polypeptides encoded by the plastome and chondriome are components of the respective chloroplast and mitochondrial proteins. The other components of these proteins are encoded in the nuclear genome. Thus, only a well- concerted production of polypeptides by the nuclear-genome and the organelle-genomes will result in organelle functionality, and consequently in whole plant functionality. Additional information on the plastome and its interaction with the nuclear genome can be obtained from the publications of M. Sugiura and collaborators (e.g. Sugiura, 1992; Wakasugi *et al.*, 1997; 1998) as well as from the review of Pyke (1999). Further information on the chondriome and its interaction with the nuclear genome is provided by Breiman and Galun (1990) and more recently by Mackenzie and McIntosh (1999). It should be noted that the size of the plastome is rather conserved in the plant kingdom, comprising about 150,000 nucleotides, while the chondriome is rather variable in size even within the same plant family.

1.1. Background: Genes and Their Expression

1.1.1. *Basic concepts concerning genes*

The term *gene* was coined by Johannsen (in 1909) as an element of inheritance, i.e. based on form and pattern. With the advancement of genetic understanding, the term *gene* was also conceived as a sequence of deoxyribonucleotides (in some cases — ribonucleotides) that code for polypeptides or RNA species (e.g. ribosomal RNA and tRNA). We shall look at genes in more detail in subsequent parts of this book but here, we should note that a *gene* comprises more than the sequence that codes for a protein or a given RNA type (the *coding sequence*). Rather, it also includes "upstream" (i.e. towards the 5′ end) as well as "downstream" sequences, flanking the coding sequence. Moreover, most eukaryotic nuclear genes (and many genes in the organelle genomes) contain, in addition to coding sequences (*exons*), also intervening sequences (*introns*), that are removed ("spliced") before the transcribed RNA is translated into polypeptides. We shall go into more detail of the structure of genes in subsequent chapters when dealing with gene expression, but for a comprehensive understanding, the reader should consult appropriate texts on molecular genetics (e.g. Alberts *et al.*, 1994; Singer and Berg, 1991).

1.1.2. *Is the living world monophyletic?*

The living world is cellular. Even when we include viruses to this world, viruses require cellular hosts for their multiplication. The living cell is always encompassed by a membrane (with or without a cell wall). It may have also a discrete nucleus. Organisms having nucleated cells are termed eukaryotes. The genetic information in the nucleus is organised in chromosomes, which are long DNA molecules that are complexed with specific proteins. It should be noted that during a certain part of the division cycle of the cells, the nuclear membrane disentangles, thus during this phase, there is no discrete nucleus and the chromosomes are rendered more condensed but are not separated from the cytosol by the nuclear membrane. There are exceptions to this dissolution of the nuclear membrane (fungal organisms). Prokaryotic organisms (bacteria) lack a discrete nucleus and have no authentic chromosomes.

Despite this variation between eukaryotes and prokaryotes, all the basic molecular-genetic entities of the cellular organisms — whether single-celled or multicellular — are very similar. Indeed, not only is each amino acid coded by a "codon" of three nucleotides but each triplet nucleotide codes for the same amino acid throughout the living world. There are a few exceptions to these codings in certain protozoa and in mitochondria, but all these exceptions seem to stem from the "editing" of the respective RNA which is the intermediary link between the coding sequence in the DNA and the resulting polypeptide. These similarities in molecular genetics and especially the universality of the coding system lead to the conclusion that all living forms have a common origin, indeed, they probably all descended from *one* archean cell.

1.1.3. *Gene flow: past and present*

The single-archean-cell possibility means that in the early epochs of evolution, there were even greater similarities between individual organisms than those existing today and thus "horizontal gene flow" did occur frequently. A theory initiated by Mereschkowsky (1910), publicised by Margulis (1981) and more recently reviewed by Gray (1993) concerning "endosymbiosis", claims that not only individual genes were "flowing" from one to another organism, but that the entire genome of one organism entered another one and this symbiotic relationship is the origin of organelles in eukaryotes. The question of the origin of eukaryotes was discussed thoroughly by Herrmann (1997). The subject of horizontal gene transfer was elaborated by us in our previous book (Galun and Breiman, 1997) and we have no way to determine the ease and frequency of this transfer in archean periods, during the evolution of the cellular world. In the present epoch, there are numerous barriers for gene transfer, even between closely related species. Even microorganisms have efficient means to recognize and destroy foreign DNA by an arsenal of endonucleases. These enzymes will cut in the recognition sites of 4, 6 or 8 specifically-ordered nucleotides that are alien to those of their own DNA and by that, destroy the alien DNA. Based on existing

DNA sequences in plants and other organisms, Rubin and Levy (personal communication) suggested that horizontal transkingdom gene flow is very rare. In angiosperms, there is the complex barrier of sexual crossability that allows sexual crosses among plant of the same species, but renders interspecific and intergeneric crosses very difficult or impossible. There are means to overcome this sexual barrier by cell manipulation such as somatic hybridization of isolated plant protoplasts and cybridization (e.g. fusion between recipient-protoplasts and donor-protoplasts; the latter with inhibited nuclear division) that may lead to plants with alien organelles (see Galun and Aviv, 1986). Such cell manipulations may transfer whole genomes rather than specific genes into the host. Even in a specific method of cell manipulation termed "asymmetric somatic hybridization", in which the donor protoplasts are exposed to a high dose of gamma irradiation — the transfer will be of chromosomal fragments rather than discrete gene. Furthermore, cell manipulation cannot bridge between organisms that are phylogenetically too far away from each other (e.g. in two different plant families). Thus, if we are interested to transfer genes into a given plant from other families and indeed from other kingdoms (e.g. bacteria, fungi or mammals), cell manipulation methods such as protoplast fusion will not lead to the anticipated results. Other methods are required to circumvent the sexual barrier and to transfer discrete genes or gene clusters into a given plant from a phylogenetically distant source. Presently, the method of choice is genetic transformation.

1.2. DNA: Chemistry, Coding and Relevance to Genetic Transformation

Providing all the knowledge on genes and their expression that accumulated since the discovery of the laws of inheritance in the monastery of Brno (by G. Mendel in 1865, experimenting with peas) and the discovery of deoxyribonucleic acid in the sperm of fishes from the Rhine (by F. Miescher in 1869) about 600 km from Brno, is much beyond the capacity of this book. But for the benefit of readers who are not versed in this subject, we shall provide a sketchy summary of this knowledge. While doing so, we shall also point out the various aspects of molecular genetics that are of specific relevance to genetic manipulation and especially to the genetic transformation of plants.

Convincing evidence for the claim that DNA contains genetic information was already provided in 1944. This was based on *transformation* experiments with the bacterium *Streptococcus pneumoniae* by Avrey *et al.* (1944). Notably, not all geneticists of the time were convinced. One of the most prominent geneticists at that time, R. Goldschmidt, in the early fifties, still insisted that DNA cannot store genetic information and attributed this storage to proteins. Ironically, Goldschmidt's detailed claims (Goldschmidt, 1955) were published after the elucidation of the structure of the double-helix DNA by Watson and Crick (1953). It took about 80 years to bridge unequivocally between the findings from the fish sperm and those of the Brno Monastery peas. The double-helix model of J. Watson and F. Crick was not only a landmark in molecular genetics, it also provides an example for the motto of this book: Watson and Crick were in an intensive race against time to come up with their (single page!) publication, worried that they will be scooped by the crystallographers R. Franklin and M. Wilkins who accumulated the diffraction information and by L. Pauling who was an outstanding specialist in the interpretation of such information. Thus, the very short *Nature* publication caused a revolution, just as a previous very short *Nature* publication by Julio-Curie on the chain reaction caused a stir in elementary particle physics.

It is by now well established that the genetic information in cellular organisms is stored in the double-helix DNA. Its fundamental chemical structure is uniform. It is composed of two strands of anti-parallel nucleic acids. The links of the strands are made of a phosphorylated sugar (deoxyribose) in which the 1′ carbon of the sugar is bound with one of four possible bases. There are two purine bases, adenine (A) and guanine (G) and two pyrimidine bases, cytosine (C) and thymine (T). The bases on the two strands are facing each other and the hydrogen bonds between A and T and between G and C are keeping the strands in their typical helical pattern. A complete helical turn occurs at about 3.4 nm and the double helix has a major and a minor groove and a diameter of about 2 nm. The "vertical" distance between any two pairs of bases is about 0.34 nm, hence there are ten base-pairs in each complete turn. But this is the "resting" pattern of the DNA. The strands separate partially during replication of the DNA and during its transcription into RNA. Illustrated details on the structure of DNA and its replication as well as on its histone packaging, the structure of chromatin and the composition of chromosomes, are provided by textbooks (e.g. Raven and Johnson, 1996; Singer and Berg, 1991).

This "resting" pattern, as described in some general biology textbooks, is appropriate for presenting a general idea of the DNA. But it is far from providing information on the actual "biological" DNA. We shall consider below some points that are relevant especially for our theme of genetic transformation. We should recall that in the eukaryotic nucleus, the DNA is not naked. Rather it is complexed with proteins (histones) to form the chromatin. The latter contains nucleosome cores, composed of histone octamers (two of each of the histones H2A, H2B, H3 and H4). The DNA is bound by one and three-quarter turns (commonly 145 nucleotides) around each nucleosome core, then continues as a string to the next nucleosome. The overall shape of this chromatin is reminiscent of beads-on-a-string. The beads may be rather packed and the whole complex is further packed again. All these require a considerable coiling of the DNA. The different bases help in this coiling. Thus, while the "average" angle between two pairs of bases is about 34 degrees, some base-pairs form a different angle such as 27 degrees or 40 degrees. This supercoiling and complexing with histones has several consequences. For example, when we deal with

Agrobacterium-mediated genetic transformation, we shall learn that at a certain stage, a single-stranded DNA from the *Agrobacterium* plasmid (the T-DNA) is integrated into the nuclear plant DNA. This integration will most probably occur in the "string" part of the DNA rather than in the DNA that is bound around the nucleosome core. It is assumed that in this core, the DNA is partially hidden by the histones. Furthermore, as noted above, the DNA is not always in the form of a double helix. For two vital processes, the two strands have to separate: during the processes of DNA replication and transcription of RNA from the DNA template. We may thus ask if the integration of the T-DNA from the *Agrobacterium* plasmid is more prevalent at sites which are in the phase of replication or the phase of transcription. There is no clear answer to this question but *Agrobacterium*-mediated integration into plant DNA seems to be more prevalent into domains of plant genes than into stretches of DNA-lacking genes (heterochromatin).

Notwithstanding the supercoiling and twisting of the double-stranded DNA and its occasional unwinding to single strands, each strand can still be viewed as a linear array of bases. This sequence of bases has at least in part significant meanings. First, there are the well-known sequences, the coding sequences, where each of three consecutive bases codes for one of twenty amino acids. There are 64 possible sequence combinations of base triplets and 60 of them are codes for amino acids ; so several different triplets may actually code for the same amino acid. The other four triplets have additional "meanings" (e.g. beginning and end of translation). But, only a minority of the total nucleotide sequences in the DNA are coding sequences. This results from a sequence of events. First, only a fraction of the nuclear DNA is transcribed into RNA. Only a part of the transcribed RNA is pre-messenger RNA (pre-mRNA). The pre-mRNA itself is processed; parts of it are cut out before the mature mRNA is trafficked from the nucleus to the cytoplasm. The percentage of coding sequences in the nuclear DNA varies between plants. We should note that the difference in DNA content per nucleus is vast. For example, in the bread-wheat nucleus, there is about 100-fold more DNA than in the nucleus of the "model plant" *Arabidopsis thaliana*. On the other hand, the amount of genetic information (i.e. the total length of coding sequences) in the nuclei of these two plant species is probably similar. These vast differences in DNA per nucleus could affect the efficiency of genetic transformation but information on these efficiencies is still not available. On the other hand, as it is reasonable to assume that in *Arabidopsis*, the genes are more densely situated than in wheat (or other cultivated plants as potato) thus, any DNA fragment that is arbitrarily integrated into the nuclear genome of *Arabidopsis* has a greater chance to enter into one of the genes of the host genome than in wheat. As such, an integration into the host coding sequence, when occurring at random, will disrupt the respective gene, the "knock out" of host genes by genetic transformation is expected to be more prevalent in host plants with small nuclear genomes than in hosts with very large nuclear genomes.

1.3. DNA Replication

A logical consequence of the double-helix structure was that this structure predicted the mode of DNA replication. Indeed, in a publication that followed the "revolutional" one, Watson and Crick suggested that for replication, the hydrogen bonds should open and the two strands should separate. Then, each of them will serve as a template for a new strand. This prediction reflected reality but reality turned out to be by far more complicated. The complication is greater in bacteria than in their circular DNA plasmids, and much greater in chromosomal DNA than in bacterial DNA. First, DNA replication occurs from the 5′ end towards the 3′ (these dented numbers are the respective carbon atoms of the deoxyriboside in the polynucleotide strand). This already means that the replication in a given stretch of the double helix is anti-parallel, as the two strands' orientations are 5′ to 3′ and 3′ to 5′, respectively. The opening of the double helix at the sites of replication origins causes *replication forks.* From the "corner" of such a fork, one strand (the one that runs from 5′ to 3′) can be replicated continuously (in the direction of the "corner", pushing the "corner" and unzipping the double helix). This is the *leading strand;* while the other strand (termed *lagging strand)* replicates in fragments, away from the "corner". The latter fragments are termed *Okazaki fragments.* Bacterial plasmids, plastomes and bacteria have usually one replication origin in their DNA. Although it is quite frequent, especially in actively dividing bacteria, that as the replication fork moves and the DNA reanneals after replication, the replication origin opens another fork so that quite a number of replication processes take place simultaneously in the same bacterial genome. In eukaryotes, there are many replication origins along each chromosome. These may be spaced out by different lengths of DNA. The space between any two of such "origins" may be 20,000 base-pairs (20 kbp) to 100 kbp. The origins do not start replicating simultaneously but in clusters. Thus, many replication processes operate at the same time. The eukaryotic chromosome has regions termed "heterochromatic". These are regions of tightly packed chromatin. These regions undergo DNA replication in the final stage of the S phase (the DNA replication phase) of the cell. In special cells, as in eggs of some animals, the whole cell cycle is very rapid. In these cells, the replication origins are spaced at intervals of only a few thousand nucleotide pairs. In such cells, there are replication origins that start replicating DNA without the requirement of a specific DNA sequence. There is a controlling mechanism which ensures that in each replication origin the replication occurs only once during a given cycle of cell division. For the initiation of replication a bubble of two single strands is opened and replication starts in both opposite directions until the replicated strands meet. It is reasonable to assume that the replication origins have unique base sequences. This assumption has relevance to genetic transformation. If an integrated alien fragment of DNA is inserted within such a unique sequence, the respective origin of replication may be impaired. We do not know if all the replication

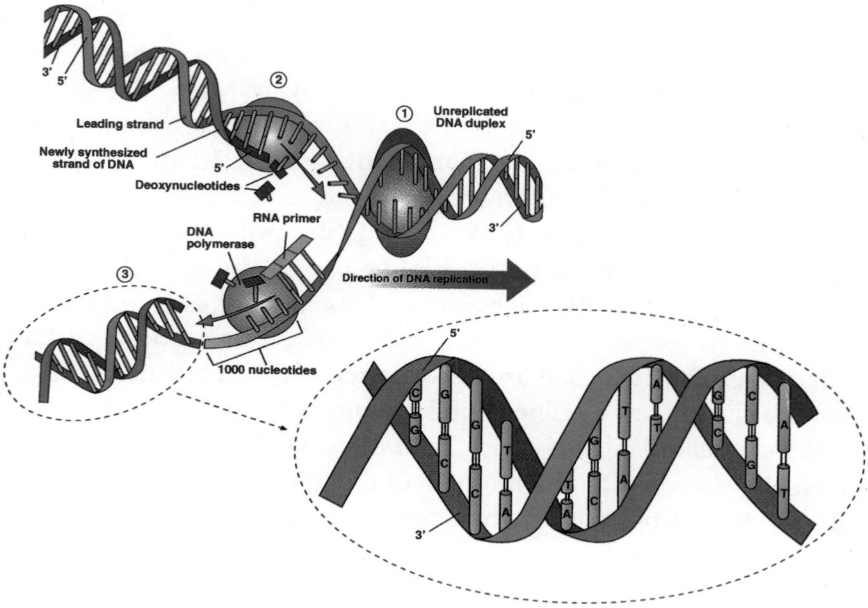

Fig. 1. A highly schematised drawing of DNA replication. (1) Unwinding by helicase; (2) Addition of nucleotides by DNA polymerase.

origins of an eukaryotic chromosome are indispensible. On the other hand, we should remember that during genetic transformation, integration is expected in many cells. Thus, the impairment of DNA replication and consequently the slowing down and even the stopping of cell division will not have a significant effect on this transformation. A scheme of DNA replication is shown in Fig. 1.

As noted above, the process of DNA replication, especially in eukaryotes, is rather elaborate. Its details in metazoa are being revealed gradually but relatively little is known about this process in plants. Let us summarize the main events and mention the enzymes that are involved. First, the replication does not happen in the condensed metaphase chromosomes but rather during the interface when the DNA is within the relatively loose chromatin. When a replication origin starts to be active, this activity should occur on the stretch of DNA that is between two nucleosomes. The opening, i.e. the separation of the two strands is caused by *helicases.* To keep the strands apart, they are stabilized by single-strand binding proteins. The addition of a 5′ deoxyribonucleotide unit to the existing 3′ hydroxyl group on the elongating strand is accomplished by one of a group of *DNA polymerases.* The DNA polymerase recognizes the origin of replication but it requires a *primer* to which it adds the incoming deoxyribonucleotides. This primer is a fragment of RNA that hybridizes to the template-DNA strand. These additions of deoxynucleotides are in accordance with legitimate binding (e.g. A to T and G to C). The legitimate bindings are favored but there are occasional mistakes. These mistakes (as A paired with C) are almost always corrected by removal of the wrong deoxyribonucleotide and its replacement with the correct one. In the *Escherichia coli* bacterium, it was found that one DNA polymerase (*Pol I*) has several activities. First, it catalyzes the addition of 5′ deoxynucleotidyl units and by that synthesizes the new strand along its template. Then, it may act as a 3′ to 5′ *exonuclease* on single-stranded DNA, removing deoxyribonucleotides from the 3′ end. There is an additional exonuclease activity of *Pol I* : removal of deoxyribonucleotides from double-stranded DNA, starting from the 5′ end. The exonuclease activities can be removed from *Pol I,* thus retaining only the DNA polymerase activity. The retained fraction of this DNA polymerase is the "Klenow fragment". The latter fragment is an essential tool in genetic engineering. It is very useful during the construction of transformation vectors. The *Pol I* is also capable of performing another activity that is important in genetic manipulation: *nick translation.* It can extend the 3′ end of a strand at a nick (a space left after a deoxyribonucleotide was removed) in a double-stranded DNA and remove deoxyribonucleotides from the 5′ end of the same nick, thus producing a replacement strand fragment. If the *Pol I* uses radioactive deoxyribonucleotides, a radioactively labeled DNA is synthesized. There are other additional DNA polymerases in *E. coli,* and other bacteria also vary from *E. coli* in their DNA polymerase arsenal. The knowledge of eukaryotic DNA polymerases is less advanced. Thus, for example, there are four DNA polymerases in mammalian cells, one of them active in the mitochondria and one is active in DNA replication in the chromosome. The role of a third DNA polymerase is in gap filling which is the addition of deoxynucleotides into gaps of the DNA from which previous deoxyribonucleotides were removed. We know very little about the DNA polymerases of plants. Moreover, there is only a small chance that knowledge on this subject will expand considerably in the near future. There are two reasons for such a forecast. First, there are much fewer investigators engaged in the molecular biology of plants than in the molecular biology of microorganisms and mammals. Then, the relatively few investigators involved in molecular biology of plants have a tendency to avoid the investigation of systems that were revealed in microorganisms and mammals because they see a great chance that they will not reveal anything that is of a real novelty. Investing great efforts to reveal an elaborate system that was fundamentally already found in other organisms is not challenging. We pointed out this deficiency in knowledge on DNA polymerases in plants but we should keep in mind that such deficiencies exist with respect to other aspects of DNA replication in plants and indeed with respect to other areas of molecular biology of plants that have relevance to the genetic transformation of plants. We shall go back to DNA replication. We handled the elongation of the leading strand. In the lagging strand, the elongation of the strand is by the Okazaki fragments that are, in mammals, 100 to 200 nucleotides long; the initiation of DNA replication in each fragment requires an RNA primer that is later removed, replaced by deoxynucleotides and the fragments of the lagging strands are joined onto a continuous strand by DNA ligases. Such ligases are also required for other tasks as the repair, at specific locations, of damaged DNA.

There is another formidable topological problem in DNA replication. The replication requires a quick rotation of the double strand around its long axis. The opening of the "negatively" coiled double strand then causes a positive coiling in the helix beyond the fork. This would cause a very complicated knot. This topological problem is greater in eukaryotes in which the double helix is complexed with histones, than in prokaryotes. This "problem" is resolved by DNA topoisomerases. These enzymes cause nicks in the coiled double strand, creating "swivels" and allowing separate rotation of the double helix beyond the fork from the nascent double strand; thus avoiding the knots and tangles that would impair the propagation of the replication fork. These enzymes also close the nicks of the affected strand.

There are two issues in DNA replication in eukaryotes that are becoming gradually clearer. One concerns the connection of a newly made DNA strand to the previous one. The former is left with a 5′ end, meaning DNA polymerase will not extend it. There are several possible ways to overcome this problem, all of them quite elaborate and far from clear in plants. Another issue is the question of what happens at the distal ends of eukaryotic chromosomes — the telomeres. Here, again, special DNA patterns such as tandem repeats and "hairpin" loops were found as well as a special group of enzymes, the telomerases. It seems that there are controlling mechanisms in eukaryotes that affect the continuity of cell division by their impact on DNA replication in the telomeres. In man, there is good evidence for a "clock" that measures the number of replications of chromosomal DNA before the cells stop further DNA replication. In embryonal cells, the length of the telomere is more than 10 kbp (the length of telomeres in individual chromosomes varies). Telomere length is reduced by the process of replication but the enzymatic activity of telomerase retards this gradual shortening of the telomeres. In the absence of telomerase activity, telomeres of some chromosomes reach a critically short length (about 1 kbp). This presumably stops further cell division (see Seding, 1998, for review). We do not know precisely how senescence in plant cells is controlled as even in mice, the senescence of cells is different from this phenomenon in man.

1.4. The Ribonucleic Acids (RNAs)

1.4.1. *The chemistry of RNA and the RNA types*

The single-stranded RNA, the form in which RNA appears in plants (but see below) is strikingly similar to single-stranded DNA, and the joining of nucleotides in RNA is identical to the joining in DNA, i.e. by the same 5′ to 3′ phosphodiester linkage. But there are two important differences. First, RNA contains the pyrimidine uridine rather than thymidine as in DNA (as well as several other modifications in its nucleotides). In addition, there is ribose in RNA with 2′-OH, while there is deoxyribose in DNA. The presence of a hydroxyl near the phosphate renders the RNA sensitive to alkali and also to RNAse enzymes.

Plant RNA is basically single stranded but, not so in many plant viruses where it is double stranded. Indeed, if double-stranded RNA is found in plant tissue, it indicates viral infection. While RNA is transcribed as single stranded, extended stretches later fold to form an apparent double strand. This is caused by base complementarity between the bases of two fragments of the same strand; the strand may fold causing pairing structures within the same strand. Thus, double-helix "stems" are formed in the paired stretch and consequently, single-stranded loops at one or both ends of these "stems" are formed. Such "stems", "loops" and "hairpins" furnish important information beyond the mere base sequence of the RNA.

Eukaryotes contain several types of RNA in their cells. Plants, in addition to having nuclei and mitochondria, also have chloroplasts. There are special types of RNA in the chloroplasts that do not exist in other eukaryotes. The chloroplast contains ribosomes (protein/RNA complexes involved in the translation of mRNA to protein) that differ from both the cytosolic and the mitochondrial ribosomes. The chloroplast ribosome, just as the cytosolic and the mitochondrial ribosomes, is built up of two subunits and each of the subunits has its own ribosomal RNA (rRNA). Another small rRNA also participates in the construction of the chloroplast ribosome. In addition, the chloroplast contains its own transfer RNAs (tRNAs). As we shall see, while dealing with translation, there are one or more tRNA(s) for each of the 20 amino acids that make up the proteins. The role of tRNAs is to bind a specific amino acid to one of its "ends" and by complementary attachment of three nucleotides from one of its loops (the anticodon triplet), to bind to a triplet of nucleotides (the "codon") in the mRNA. The tRNAs are about 70 nucleotides long and because of many stretches that are complementary to other stretches on the same (basically linear) polynucleotide, "stems" and "loops" are formed, giving the tRNA a pattern reminiscent of a clover leaf (when drawn in two dimensions). The rRNAs also contain "loops" and "stems" but they are more than ten-fold larger than tRNAs. They thus have more numerous "loops" and "stems" even though they are also composed of one (long) single-stranded RNA. The rRNAs of chloroplasts as well as the tRNAs for all the 20 amino acids are coded by the chloroplast DNA (Wakasugi *et al.*, 1998). The chloroplasts as well as the mitochondria have their own machinery for the translation of mRNA into proteins, although only a minority of the mitochondrial polypeptides are translated within this organelle. The majority of the polypeptides are trafficked from the cytosol into the mitochondrion. Many of the mitochondrial enzymes are thus made up of subunits, part of which are synthesized in this organelle and part of which in the cytosol. This, as we shall see later, has important consequences with respect to the compatibility between nuclear gene products and products of genes in the mitochondrial genome (see Breiman and Galun, 1990). Moreover, while all the rRNAs of plant mitochondria are coded by the chondrion, only part of the tRNAs of this organelle are coded by the chondrion; the other tRNAs have to enter it from the cytosol.

The nucleus of plant cells encodes three rRNAs as well as many tRNAs. There is one tRNA for each of the amino acids methionine and tryptophan, respectively, and two or more tRNAs for each of the remaining 18 amino acids. RNA is transcribed from the three genomes of the plant cells: the nuclear genome, the plastome and the chondrion. In all these cases, the single-stranded transcribed RNA is processed and the respective pre-mRNAs are produced. The mRNAs of the nucleus exit to the cytosol before performing in the translation of proteins while the mRNAs of the organelles are retained in their respective organelles.

There are additional types of RNA. One of these, that serves as primer for the replication of DNA, was mentioned before. Other types of RNA are less defined but have terms such as heterologous nuclear RNA (*hnRNA*), small cytoplasmic RNA (*scRNA*) and small nuclear RNA (*snRNA*) that were revealed in eukaryotes.

There is thus an abundance of RNA types. They serve in a variety of functions. Some of them have structural tasks (rRNAs), others are mediators (e.g. tRNAs, primers) while some have catalytic activity (e.g. in the mitochondria) similar to the catalytic function of protein enzymes. Taken together, this multitude of functions led to a hypothesis that there was an "RNA world" that preceded the present DNA–RNA–Protein world. Turning back to present reality, plant cells store all their stable information in DNA rather than in RNA. Thus RNAs, in contrast to DNAs, do not replicate in eukaryotic cells and all RNAs are transcribed from DNA templates. They may then be processed by various means.

1.4.2. *Transcription of DNA into RNA*

By *transcription*, we mean the synthesis of a single-stranded RNA from a DNA template. There are some obvious similarities between DNA replication and RNA transcription. The most obvious similarity is that, in both cases, one of the DNA's single strands serves as template for the new strand and that the bases of the incoming nucleotides are complementary with the bases of the DNA-template. Thus, the C, G and A ribonucleotides for the RNA strand are paired with the respective G, C and T deoxynucleotides of the DNA strand. But the U nucleotide of the synthesized RNA strand is paired with the A deoxynucleotide of the DNA because in RNA, uracil (U) substitutes thymine (T) of the DNA.

As already pointed out before for other molecular genetic processes, our knowledge of the transcription process in plant nuclei is relatively meager. Early studies on the transcription of DNA templates into RNA were performed with prokaryotes (i.e. bacteria) where there is only one DNA-dependent RNA polymerase that acts in this transcription. Numerous studies caused a good understanding of this process in prokaryotes. Elucidation of transcription in eukaryotes was more complicated, but intensive studies with organisms such as yeasts, flies and mammals yielded a fair understanding of this process in several eukaryotes. As for plants — again, much fewer studies and consequently much less knowledge are available. Moreover, there is no reliable textbook on plant molecular biology to which we can refer our readers. Although several studies show similarity in transcription between plants and animals (see Katagiri and Chua, 1992; Nikolov *et al.*, 1992; Ramachandran *et al.*, 1994; Meshi and Iwabuchi, 1995). Therefore, for the purpose of our deliberations, we shall assume that the transcription of RNA in the plant nuclei is similar to that in the mammalian nuclei, and thus, we summarize this transcription below.

First, let us have a second look at a typical eukaryotic gene that codes for a protein. We shall describe the gene in the conventional diagrammatic manner as a linear array of two DNA strands and neglect the double helix, the supercoiling and its complexion with histones. The "upper" strand shall be oriented 5′ to 3′ from left to right and will be termed the *sense* strand and its sequence is identical to the sequence of the transcribed RNA. The other ("lower" strand) is the *template* strand and it points to the opposite (3′ to 5′ from left to right) direction. There is one reference point: the initiation of RNA transcription. The deoxynucleotide at this point will get a number +1. The deoxynucleotides to the right (downstream) of this point will get sequential positive numbers. Upstream of +1 are the non-transcribed deoxynucleotides of the gene. Their numbers will be negative, with –1 adjacent to the initiation of transcription (+1). Looking to the left of +1, there is first the promoter sequence and to the left of it are upstream regulatory regions. The upstream border of the gene is not well defined because there may be enhancer regions for this gene that are located more than 1 kbp upstream of the initiation of transcription. Such enhancer elements may also be located in introns or in the downstream region beyond the coding region. There may also be nuclear-attachment regions (NARs) that are further upstream of the initiation of transcription. To the right (downstream) of the initiation of RNA transcription, there is first a 5′ *leader* region that is transcribed but not translated to protein. To the right of the leader is the translated region and finally, there is again a transcribed but not translated 3′ trailer region. The eukaryotic genes are divided into three classes. Each class is transcribed by its respective RNA polymerase. Of these, class I and class II genes code for a variety of RNA species (e.g. ribosomal RNAs and nuclear RNAs) which are not translated into proteins. For our theme, we shall focus on class III genes from which all mRNAs are transcribed. Clearly, the other classes are vital to the eukaryotic cell functions and it is estimated that about half of the cell transcriptions are from class I genes. But, in this book, we are dealing with proteins coded by alien ("trans") genes or with products synthesized by such proteins (i.e. such as alien enzymes) and these are all coded by class III genes. Class III genes encode, in addition to all mRNAs, also small nuclear RNPs. Again, we shall neglect the latter subclass.

1.4.3. *Transcription by polymerase II*

The transcription of mRNAs by RNA polymerase II takes place in the nucleus. Each group of eukaryotic organisms (e.g. yeasts, fungi, amoeba, insects, amphibians, mammals, plants) has its own specific type of RNA polymerase II, but the differences are minor. This polymerase is rather big and complicated. It consists of nine or ten subunits. Two of these are large, about 200 and 140 kilodalton (kDa), respectively. These two subunits are very similar to subunits in the other eukaryotic RNA polymerases and are also similar to the subunits of polymerases I and III. Thus, RNA polymerase II is unique in its remaining subunits. The fungal toxin, α-amanitin, blocks RNA elongation when applied in very low doses during transcription by polymerase II, probably by binding to its largest subunit, while the other polymerases are much less sensitive. There is an abundance of RNA polymerases in the nucleus. They survey the DNA and when they reach a given region, upstream and close to the site of the initiation of transcription, they bind to it. The bound polymerase is then assisted by several additional components.

The attachment of the polymerase to the DNA starts the build-up of a whole "transcription factory". This "factory" contains numerous *transcription factors*, many of the which are DNA-binding proteins that recognize DNA elements at the initiation of transcription where the RNA polymerase is located. Other factors have protein–protein binding capabilities. In total, this complex of protein factors regulate the initiation of transcription by the polymerase. The factors may have a positive effect, i.e. to facilitate transcription, or a negative one, i.e. to suppress transcription. Thus, the DNA motifs and the availability of factors regulate transcription. Regulation is also imposed by DNA motifs outside of the site of initial binding of the polymerase (which by itself is not precisely delimited). There are *cis* elements that can interact with *trans*-acting (protein) factors. While it is clear that the whole regulatory mechanism that regulates transcription by polymerase II is very intricate, the system was clarified in only specific cases (i.e. with specific genes in specific organisms). Briefly, after the attachment of the "transcription factory", the double-stranded DNA opens and an RNA strand is copied from the template single-strand DNA. In a short region, a double helix of DNA/RNA is formed but as the "machine" moves "downstream". The 5′ end of the newly synthesized single-strand RNA is detached and the detached strand elongates and is being further processed. The RNA polymerase moves on and after it passes the template DNA, it reunites with the sense

DNA single strand, forming again the DNA/DNA double helix. The RNA polymerase continues its movement (and RNA synthesis) until it meets a region in the DNA that causes it to be released from the DNA. The RNA is thus also released. A scheme of the transcription is shown in Fig. 2.

1.4.4. *Processing of the mRNA*

The processing of the newly synthesized single-stranded RNA transcript starts right after its release from the RNA polymerase complex. There are three main types of processings that convert this transcript into mature mRNA: (1) capping of the 5′ end; (2) removal of specific stretches (introns) from the RNA strand; and (3) removal of a fragment, at a certain distance from its 3′ end and its replacement with a polyadenosyl (poly-A) tail (see Fig. 2).

After this processing, the "gene", in its narrow sense (i.e. the coding sequence), is represented in the mRNA by a sequence of nucleotides, starting with the three bases (AUG) that code for methionine (in almost all cases) and ending with the three bases that signal the "stop" of translation (UAA, UAG or UGA). At this point, we shall note the relevance of this processing for the genetic transformation of plants. Because the coding system is identical for all eukaryotic nuclei (and chloroplasts) — a given plant will accept the coding sequence from any other organism, provided that the alien DNA contains the template for an RNA transcript that also contains non-translated downstream and upstream sequences which are well recognized by this plant. Thus, as we shall see in later chapters, that a transformation vector should be constructed in a way that proper upstream ("promoter") and downstream ("terminator") DNA sequences that are accepted by the host plant, are flanking the alien ("trans") coding region. One additional remark concerns genetic transformation. Although the coding for amino acids is universal, each amino acid (with the exception of methionine and tryptophan) can be coded by more than one codon. It was found that each organism

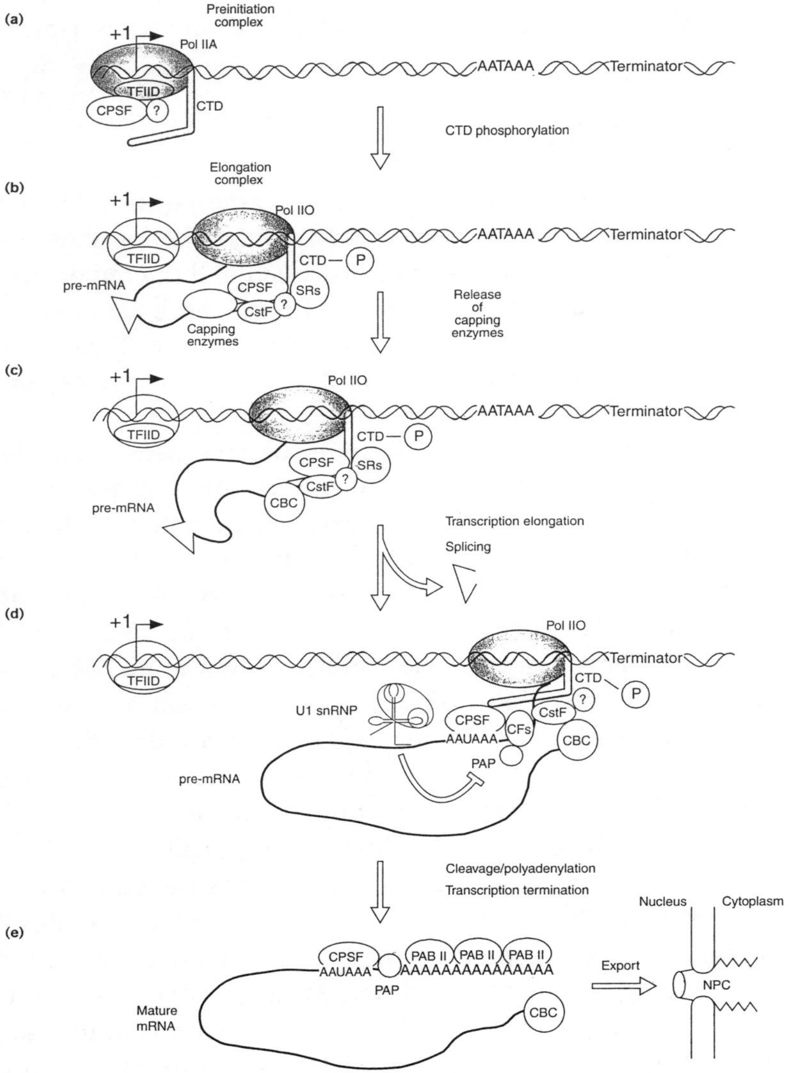
(a)
Preinitiation complex
+1
Pol IIA
AATAAA
Terminator
TFIID
CPSF
?
CTD
CTD phosphorylation
(b)
Elongation complex
+1
Pol IIO
AATAAA
Terminator
TFIID
CTD
P
pre-mRNA
CPSF
SRs
CstF
?
Capping enzymes
Release of capping enzymes
(c)
+1
Pol IIO
AATAAA
Terminator
TFIID
CTD
P
CPSF
SRs
CstF
?
CBC
pre-mRNA
Transcription elongation
Splicing
(d)
+1
Pol IIO
Terminator
TFIID
CTD
?
P
U1 snRNP
CPSF
CstF
CFs
AAUAAA
CBC
PAP
pre-mRNA
Cleavage/polyadenylation
Transcription termination
(e)
Nucleus
Cytoplasm
CPSF
PAB II
PAB II
PAB II
AAUAAA
AAAAAAAAAAAAAAA
PAP
Export
NPC
Mature mRNA
CBC

Fig. 2. Scheme of the transcription and coupling between RNA processing factors and the transcription apparatus. (a) Early during formation of the preinitiation complex, TFIID associates at the transcription start site with the 3′ end processing factor CPSF (cleavage and polyadenylation specificity factor) and perhaps other factors(?). The carboxy-terminal domain (CTD) of RNA polymerase II is unphosphorylated at this stage (*Pol IIA*). (b) Just after transcription starts, the CTD is phosphorylated (*Pol IIO*) and CPSF and CstF are found associated with it. The enzymes responsible for capping of the pre-mRNA 5′-end are also found associated with the phosphorylated CTD. This early association explains why capping occurs shortly after transcription initiation. It has also been shown that the CTD may influence splicing which could be due to the binding of splicing factors (such as SR proteins) to the CTD. (c) The cap-binding complex (CBC) bound to the 5′ cap has been shown to directly associate with the 3′-end processing a factor at the poly-A site and enhance the cleavage reaction. It is proposed that this association occurs during elongation of the pre-mRNA before the 3′ processing factors are transferred from the CTD to the poly-A site. This interaction could also account for the active role of CBC in splicing. (d) After transcription of the 3′-end processing signals, the cleavage factors CPSF and CstF may be transferred to the elongating pre-mRNA from *Pol IIO*; however, *Pol IIO* might not dissociate from the cleavage complex but remain bound and participate in the formation of a stable and active complex. Cleavage could thus represent a signal for the polymerase to terminate. As mentioned before, the CBC-cap complex suggests that the pre-mRNA is synthesized as a circular molecule. At that stage, the U1A and U1 70K protein of the U1 snRNP can regulate polyadenylation by inhibiting PAP activity. The inhibition occurs by binding to an inhibitory element located on the pre-mRNA just upstream of the AAUAAA signal. (e) Following cleavage, polyadenylation of the pre-mRNA does not require the association of the CBC or the *Pol II* CTD; however, the CBC — and most probably other proteins that bind to the mature mRNA — plays an active role in the nucleocytoplasmic export of the mRNA through the nuclear pore complex (NPC). PAP is poly(A) polymerase and PABII is poly(A) binding protein. (From Minivelle-Sebastia and Keller, 1999; reprinted from *Current Opinions in Cell Biology* **11**: 352–357. Copyright © 1999 with permission from Elsevier Science.)

has its own favored codons for specific amino acids. Thus, using an alien coding sequence (e.g. of bacteria or protozoa) in the genetic transformation of plants could retard the synthesis of the respective alien protein in the transgenic plants. This can be corrected by the synthesis of an appropriate coding sequence.

1.4.4.1. *Capping*

With the exception of a number of viruses (e.g. polio, hepatitis A virus), all pre-mRNAs undergo capping. The capping takes place in the RNA polymerase II transcripts after a sequence of about 30 nucleotides of the single-stranded RNA is released from the polymerase, i.e. at its 5′ end. During capping, the terminal phosphate from the 5′-triphosphate proximal-end pre-mRNA is removed and enzymatically linked to the guanylyl group of the nucleotide guanyl-tri-phosphate. In this process, the guanyl loses two phosphates. Further changes are imposed by enzymatic methylation of the guanosine.

1.4.4.2. *Removal of nucleotide stretches from the "middle" of the pre-mRNA*

The discovery that during mRNA maturation, nucleotide stretches are removed from newly transcribed RNA is relatively recent (1977). It was found that usually the "coding region" of the DNA is considerably longer than the respective mRNA. Sequencing clearly indicated that genes (i.e. the DNA) contain deoxyribonucleotide sequences that were not represented in the nucleotide sequences of the mRNA. Early evidence for this also came from electron microscopy studies in which the DNA of a given gene (e.g. coding for γ-globin) was spread out and annealed with the mRNA of the same gene. The pattern was analyzed, and revealed stretches of double-stranded DNA, single-stranded RNA and single-stranded RNA attached to DNA. The picture that emerged was loops of DNA that protruded from a line of mRNA bound to the DNA. The interpretation was that the mRNA hybridized only to certain sequences of the DNA. Further studies made it clear that during maturation, fragments of the pre-mRNA are removed. The removed fragments were termed *introns* and the retained fragments were termed *exons*. After removal, the ends of the RNA nucleotide sequence are joined. The process was thus termed *RNA splicing*.

How exactly components of the nucleus recognize the borders between exons and introns in order to cut out the introns in perfect precision is still enigmatic. Some "consensus" border nucleotides were reported (e.g. AC and GT, respectively, at the 5′ and 3′ end of exons). But there are many exceptions and obviously, the majority of these "consensus" nucleotides are not important in splicing. Therefore, given a DNA fragment that contains the coding sequence, we may have a good guess where the splicing will occur but cannot predict it unequivocally. In practice, it is easy to find out. When the mRNA is transcribed in reverse into cDNA, the latter's sequence can be compared with the respective DNA sequence by a rather simple computer program. When aligned, the DNA sequences that are not matched by cDNA sequences are the introns. We shall not go into detail how the splicing takes place but note that the pre-mRNA (also termed *primary* mRNA), right after synthesis, is complexed with proteins in a manner reminiscent of the complexing of DNA with histones in the chromatin. Particles of protein wrapped by RNA are formed (that are larger than nucleosomes) and these particles are connected by the continuous RNA strand. This configuration concept allows the recognition of splicing sites. Although the details are not yet known, a name was given to the entities that perform the splicing: *splicesosomes*. An update on splicing of mRNA in plants was provided by Simpson and Filipowicz (1996). Again, for the expression of alien genes in transgenic plants, it is essential that the splicing recognition of the host plant is identical to the recognition in the organism from where the gene was obtained. Obviously, no such problems arise when the transgene is derived from a prokaryote (that is not an archaebacterium) where there are no introns and exons (archaebacteria do have introns). Finally, we shall note that not all eukaryotic genes that code for proteins contain introns. The genes coding for the five histones always lack introns. Introns are also absent in most fly genes that code for proteins. But the "rules" for the existence (or absence) of introns are not known. Two species of the same genus of yeast may differ in this respect. In plants, most genes do contain introns but the lengths of introns, in the same gene but of different plant species, may differ considerably. Interestingly, while the lengths vary, the number of introns may be the same for a given gene, even in different plant species. In practice, all the above probably do not affect plant transformation because for this transformation, the respective cDNA, rather than the genomic DNA that codes for it, is used in the construction of the transformation vectors. The word "probably" should be taken seriously. It is possible that the splicing of the mRNA serves an important purpose. It may slow down the maturation of mRNA. This slow down could be vital in order to allow for the relatively slow process of correct folding of the translated protein. Sometimes, "haste is from the devil" (a Beduin maxim).

1.4.4.3. *The poly-A tail*

RNA polymerase II transcripts have beyond the coding regions a signal that marks the later synthesis of a polyadenylic-acid tail. This tail can reach a length of 100 or more adenyl nucleotides. Notably, mitochondrial transcripts do not have such tails. Chloroplast transcripts may be polyadenylated but then the poly-A tail may serve as a degradation signal (Schuster *et al.*, 1999). Thus, while it is a simple procedure to catch the tails (by running an extract of cellular RNA through a polydeoxythymidine column) and isolate the nuclear mRNA, it is difficult to fish out the tailless mRNA from the organelles. The tail is a modification that is not formed at the very (3′) end of the primary mRNA, when the RNA polymerase II as well as the RNA strand are released from the template DNA. The addition of the tail happens way upstream at a cleavage signaled by a specific motif (AAUAAA in mammals and in at least some plant pre-mRNAs). Cleavage then takes place some nucleotides downstream of this signal (e.g. 10 to 30 nucleotides). It is evident that this signal alone cannot be sufficient for the cleavage because such a sequence is expected to appear quite frequently in the code (for example, the codon for asparagine followed by the codon for lysine are: AAUAAA). Thus, there should be at least an additional preceding motif to signal cleavage. The RNA polymerase continues its transcription for many nucleotides beyond the cleavage signal (in some cases, for about 1,000 nucleotides). Then, the polymerase as well as the transcribed RNA strand are detached. The RNA strand is then progressively degraded between the 3′ end and the cleavage site, and a long poly-A is attached to the 3′ end of the RNA chain by a specific polymerase. The above description is a gross generalization. First, the cleavage and the poly-A signal can be bypassed. Thus, even the same gene may have two transcription products with mature mRNA of different lengths. It also seems that T-rich sequences tend to loosen the attachment of the polymerase to the DNA template; hence, to assure continuous attachment, an "anti-termination" factor (elongation factor) is attached to the polymerase and only upon its separation from the polymerase, the latter can be detached from the template and terminate transcription. Several roles were attributed to the poly-A tail: (1) it aids the export, from the nucleus of the mature mRNAs; (2) it improves the stability of the mRNA in the cytosol; (3) it accelerates the loading of the mRNA onto the ribosomes during translation; (4) as shall be noted in the section dealing with translation, at least in plants, there is probably a binding between the 5′ cap and the poly-A tails. But the mature mRNA of animal histones lack poly-A tails, while in plants, the mRNAs of histones have such a tail. Only RNA transcripts that are synthesized by RNA polymerase II are capped and attain poly-A tails.

Therefore, if by recombinant DNA methods the cDNA for a given eukaryotic protein is attached behind a promoter that is recognized by the RNA polymerase I or RNA polymerase III, and this construct is used in genetic transformation, the resulting mature mRNA will be neither capped nor polyadenylated. Such a mRNA may not be operational in the host plant.

1.4.4.4. *Primary transcript versus mature mRNA in the cytosol*

The amount of mRNA that is exported from the nucleus reaches only about 1/20 of the total transcribed RNA. So, the vast majority is "lost" in the nucleus. This seems as a great waste to be conserved during evolution of higher eukaryotes (e.g. mammals). Can we surmise that energy saving does not play a decisive role in evolution at either the cellular level or at the whole-organism (human) level?

1.5. Translation of mRNA into Protein

An extensive number of components collaborate in the translation of proteins from mRNAs. These include the *transfer RNAs* (tRNAs) and the ribosomes that are organelles composed of an interaction between ribosomal RNA (rRNA) and ribosomal proteins. In addition to these are two types of translation factors (proteins): about a dozen initiation factors and several elongation factors. Finally, an extensive number of enzymes are involved in translation.

These components encompass an elaborate machinery that must operate in perfect concert. Malfunction in one of the components may abolish the protein-synthesis capability or retard it. However some components such as the ribosomal RNAs are produced by many genes of the same genome, so that if one gene is not functional, the damage will be compensated by other genes. We shall see below that genetic transformation can cause the insertion of the transgene into a gene of the host plant. Such an insertion can abolish the activity of the respective host gene. Thus, a machinery coded by 100 or more genes is vulnerable to the insertion of transgenes. We should consider that plant genetic transformation is affected by hundreds or thousands of cells in each operation. Impairment of translation in a given cell will prevent its further divisions. But, other cells with an intact translation activity can take-over. Therefore, any cell line or regenerated plant that survived the transformation should have an operational protein-synthesis activity.

There are some good and relatively recent reviews on the subject of translation in plants. These can serve those interested in the details of the translation apparatus in plants (e.g. Browning, 1996; Futter and Hohn, 1996; Gallie, 1996; Bailey-Serres, 1999).

Before dealing with the components and processes of translation, we shall very briefly describe the main events of this process. The mature mRNAs are trafficked from the nucleus, through the nuclear pores into the cytosol. There, the small subunit of the ribosome binds a methionine-tRNA ($tRNA^{met}$) to which a methionine is attached. The small subunit ribosome with its $tRNA^{met}$ and initiation factors binds via its anticodon to the first triplet of the coding region (AUG) of the mRNA. Additional initiation factors and the large subunit of the ribosome join the complex and the actual translation starts with the addition of elongation factors and another tRNA, with its respective amino acid, that binds to the ribosome/mRNA complex, adjacent to the $tRNA^{met}$. The amino acid of the incoming tRNA is attached by a peptide bond to the methionine on the $tRNA^{met}$ and then this methionine is detached from the $tRNA^{met}$. There is a shift in places: the $tRNA^{met}$ is removed from the ribosome and its place is taken by the next incoming tRNA. This process is repeated until the translation complex (ribosome/elongation factors) reaches a site on the mRNA that has a "stop" codon. The translation complex is then detached from the mRNA with the help of *release factors*.

The newly formed chain of amino acids (the nascent polypeptide) folds into an appropriate three-dimensional structure at a very early stage; usually much before its translation is finished. When the terminal end of the polypeptide is released from the mRNA, the latter can serve in another round of translation.

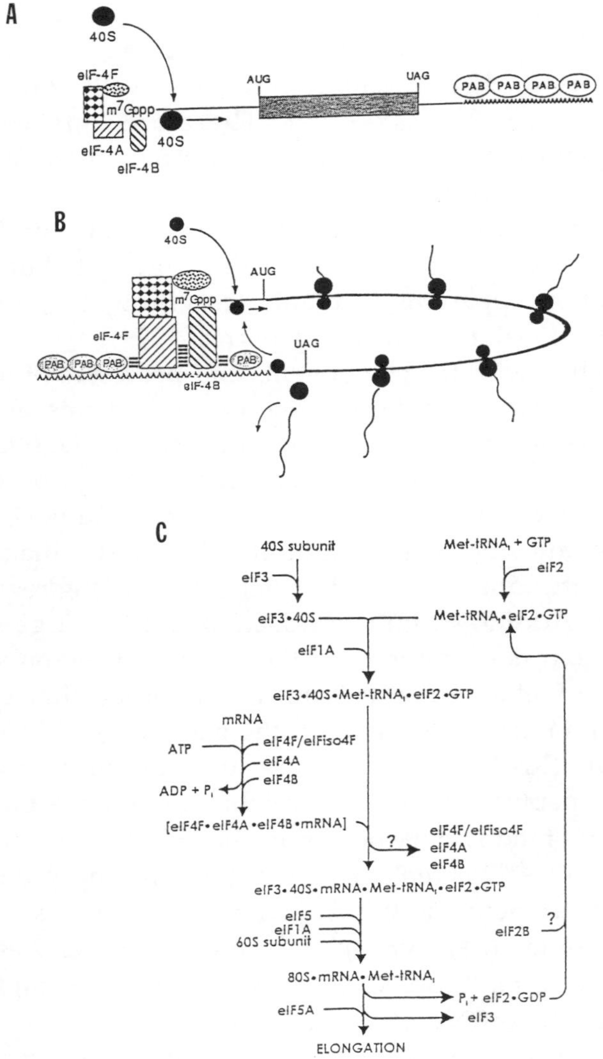

Fig. 3. Scheme of the intermediate steps in the initiation of protein synthesis in eukaryotes. eIF1A, eIF2 etc., are elongation factors; PAB is poly-A binding protein. (A) Diagram of a mRNA and proteins that bind to its 5′ terminal cap and its poly-A tail (from Gallie, 1996). (B) The codependent model of translation. (From Gallie, 1996.) (C) Scheme of complexing involving mRNA, ribosomes and elongation factors. (From Browning, 1996.)

1.5.1. *Initiation of translation*

Our understanding of the initiation of translation in plants improved considerably during the last 15 years. The details are much beyond what can be described in this book. Our summary will mainly be based on reviews by Browning (1996), Futter and Hohn (1996) and Bailey-Serres (1999). We provide this summary to demonstrate the complexity of the initiation of translation. This initiation has relevance for the genetic transformation of plants and should be taken into account before constructing transformation vectors. An overview on transcription is outlined in Fig. 3 where the involvement of initiation factors is schematically presented.

Before initiation takes place, the small (40S) ribosomal subunit combines with the factor eIF3. The latter factor is composed of ten subunits in wheat germ. It is probably essential for the subsequent binding of the eIF2•GTP•met-tRNA^{met}•40S to the mRNA. In parallel, events are taking place at the mRNA. The first step is the binding of one factor, eIF4F, to the 5′ cap structure (m^7 GpppN, where N is any nucleotide). This binding is followed by two additional factors: eIF4A and eIF4B and together with energy from ATP, a complex is formed that is instrumental to "open" any secondary mRNA structures (i.e. stem-loop structures, downstream of the translational initiation) that could prevent the scanning of the initiation complex along the mRNA. The initiation complex is further shaped by the joining of the 40S ribosomal subunit to which eIF2, GTP, and tRNA^{met}, with its attached methionine, are bound. The complex is attached very near to the 5′ end of the mRNA. An unexpected synergism was discovered between the cap and the poly-A tail of the mRNA. A fully functional translation requires the presence of both ends of the mRNA. Moreover, the two factors, eIF4F and eIF4B that bind to the cap area, also show specific and strong binding to the poly-A tail (see Gallie, 1996, for review).

1.5.2. *Elongation of the translated polypeptide*

When the initiation complex with the 40S subunit ribosome reaches the start codon (AUG), it is joined by the 60S ribosomal subunit and the full-size ribosome is established. This complex already includes the met-tRNA with its attached methionine. From here the complex has to move along the mRNA with strides that are three nucleotides long. The first factor involved in this process of incoming aminocyl-tRNAs is the elF1. In this process, the incoming aminoacyl-tRNA is placed in the correct site in the elongation complex (containing now the whole ribosome) and is pairing with its respective anticodon to establish close proximity between the 5′ terminal cap and the poly-A tail. A long mRNA loop could be formed that enables the same 40S subunit, when it reaches the "stop" codon, to proceed directly to the cap and be recycled. There is another protein, the poly-A-binding (PAB) protein that is attached along the poly-A and probably helps to keep the termini-binding (eIF4F and eIF4B) factors attached to the poly-A.

The next stage in elongation is the translocation of the tRNA. In the beginning of translation, the $tRNA^{met}$ moves out of the translation complex and the tRNA next to it replaces its site. This vacates a site for the third aminoacyl-tRNA. The translocation continues and each time, the amino acid on the incoming tRNA establishes a peptide bond with the last amino acid of the growing polypeptide. The translocation requires another (rather large) elongation factor: elF2. A detailed description of this process as well as the elongation factors involved in it are provided by Browning (1996). One additional note concerns the secondary structure of the mRNA. Due to "internal" base pairing, "stem-loop" structures and other structures can be created in the otherwise linear mRNA. These create a potential hurdle for the translation machinery. It appears that when the whole complex is established and moves beyond the first codon on the mRNA, the machinery has the capability to "open" these structures and move along the coding nucleotides of the mRNA. But before reaching the initial AUG triplet, i.e. in the "leader" sequence between the cap and AUG, the scanning of the translational machinery must overcome such hurdles. This is where the initiation factors play a role and the specific sequence of the leader as well as its length are decisive for swift scanning. For example, if the distance between the cap and AUG is below a given number of nucleotides — there is not enough length for the initiation factor to "open" the hurdles on the leader. The result is that initiation (and consequently translation) is slowed down. This "leader" sequence may therefore be an important consideration when an alien gene is to be efficiently translated in a transgenic plant.

The initiation factors noted above, as well as others that may be revealed in the future, have thus an important role in the regulation of translation of mRNAs. If the plant regulates the availability of these factors, it will regulate translation because it was observed that a shortage in these factors slows down translation. On the other hand, "smart" plant viruses and investigators may take advantage of this situation. If a leader of a given mRNA is constructed in a manner that is less dependent of such factors, then this mRNA may be preferably translated during a shortage of these factors (during stress?). The initiation factors may also have a role (e.g. change in phosphorylation) at different developmental stages of the plant or in certain environmental conditions. Again, such changes in the initiation factors could regulate translation. The phosphorylation of one ribosomal protein, the S6 protein, also seems to play a role in initiation efficiency. These as well as other considerations involved in translational efficiency were reviewed by Bailey-Serres (1999), and the optimization of alien mRNA in transgenic plants was discussed by Koziel *et al.* (1996). The latter authors also provided a possible consensus sequence around the AUG codon for optimal translation initiation in plants: UAAACAAUGGCU.

1.5.3. *Termination*

During elongation, the two tRNA attachment locations of the ribosome are occupied. One, the P-location, is occupied by the tRNA that has the polypeptide chain attached to it. The other, the A-location (conventionally written on the right side), is occupied by the incoming tRNA that carries the next amino acid. When the ribosome reaches a termination ("stop") codon (UAA, UAG or UGA) in its A-location, another factor goes into action: a *release factor*. The latter binds to the ribosome and the peptidyl-tRNA is hydrolyzed. The nascent polypeptide is released from the translation complex and the uncharged tRNA is also released. The ribosome dissociates into its two subunits, one of which, the 40S subunit, can be recycled on the same mRNA, ready to participate in the initiation of translation again. The release process, in prokaryotes and in the investigated eukaryotes, requires hydrolysis of GTP. While this brief description of termination was verified for some eukaryotes (yeast, rabbit), the termination process in plants is still not elucidated.

1.6. Trafficking and Modification of Proteins

Proteins are either the final products that we handle in this book, such as antibodies, antigens, therapeutic peptides and enzymes, or they are involved as specific enzymes in the production of high-value compounds (e.g. steroidal alkaloids). While initially synthesized as linear polypeptides, they are then folded specifically, undergo maturation, and are trafficked within and out of cell organelles (e.g. endoplasmic reticulum, Golgi apparatus, vacuoles, mitochondria, chloroplasts), or are moved out of the cells. The description of the modifications that the nascent polypeptides undergo and their various routes in and out of cell organelles is beyond what we can provide in this book. On the other hand, the investigator who plans the construction of a chimeric gene in order to harvest a given protein product in transgenic plants should not neglect several important issues. For example, plant proteins (as proteins of other organisms) have amino- and carboxyl-end signals that direct their passage in and out of organelles, such as the endoplasmic reticulum. These signals may be cleaved off after the respective passage. In some cases, there is more than one signal and each is removed in turn after the respective passage. Proteins are also modified by glycosylation and other chemical modifications. A special group of proteins, *"chaperones"*, are instrumental in maintaining the correct folding of newly synthesized proteins. One of the early findings can serve as examples. Insulin is synthesized as a linear polypeptide. It is then folded and the folded structure is still maintained as prepro-insulin by two bisulfide bonds. Then, an amino-end peptide chain is cleaved off to result in pro-insulin. The latter is again cleaved to become the final active insulin. Plant proteins can be packed (as storage proteins in seeds and grains) in specific protein bodies. We shall refer to these issues when we deal with specific products of transgenic plants. For a general overview of the fate of proteins after their synthesis by the translation complex, as well as other subjects concerning protein interactions, trafficking and folding — a general text on cell biology is recommended (e.g. Alberts *et al.*, 1994). In plants, the routing of proteins was studied quite intensely since many years ago (see Morre and Mollenhauer, 1971, for an early review) and several reviews appeared more recently: Denecke *et al.* (1990), Koziel *et al.* (1996), Galili *et al.* (1998), Herman and Larkins (1999) and Vitale and Denecke (1999). Very briefly, proteins that are destined to stay in the cytosol and go to chloroplasts and mitochondria are synthesized on free ribosomes. If they have to be targeted to specific locations, i.e. to the intermembrane space of the mitochondria, they have special signal peptides and after their arrival, part of the peptide is cleaved off to yield the "mature" protein. Proteins that are destined to end up in the vacuoles, Golgi apparatus, protein bodies, or to be secreted out of the cell are synthesized on ribosomes associated with the endoplasmic reticulum (ER). Actually they start to enter the lumen of the ER much before the full-length polypeptide is synthesized. Also, the appropriate folding may start soon after the release of a short polypeptide from the translation machinery. When the amino end of a nascent chain comes out of the translation machinery that is associated with the ER, the crossing of the ER membrane is facilitated by a factor: signal sequence recognition particle (SRP) that recognizes the amino-end signal of the nascent chain. After the chain enters the lumen, the amino end may be anchored to the outer ER membrane until the full-length polypeptide enters the lumen. Then, the anchor is cleaved off and the polypeptide starts its Odyssean voyage in the ER and beyond. A multitude of "adventures" can happen to the protein during this voyage in the ER but apparently, the protein is capable of keeping its "travel ticket" and respectively, even after modification it is directed into its proper destinations (e.g. Golgi apparatus, vacuole, outside of the cell).

Chapter 2

Genetic Transformation

2.1. Background

The theme of this book is focused on specific aims of plant genetic transformation, as indicated by its title. We shall thus avoid the handling of what is still up to now the main purpose of this transformation: crop improvement. Consequently, this chapter will handle the methodologies of plant genetic transformation in a rather different manner than in the previous book (Galun and Breiman, 1997). We shall avoid here the history of transfer and expression of alien genes in plants. On the other hand, we previously neglected methods that lead to transient expression of alien genes. Because such an expression is useful for the manufacture of some medical products (e.g. antibodies), we shall discuss these methods in this chapter. Also, in the previous book we reviewed transformation mediated by *Agrobacterium tumefaciens* but not by *A. rhizogenes*. In this chapter, we shall also describe the methods of *A. rhizogenes* transformation because this transformation is useful for the accumulation of medical products in plant roots.

There are interesting and promising innovations in genetic transformation of the chloroplast genome (the plastome), we shall therefore update this transformation in this chapter. But, most of our attention will be directed to the two important transformation methods: *A. tumefaciens*-mediated transformation and biolistic methods.

Improvements in understanding the molecular regulation of gene expression, efficient procedures to check the results of transformation as well as advances in transformation procedures led to a situation that an almost unlimited number of aims can be achieved by plant genetic transformation. This brings us to a saying attributed to Zeev (Vladimir) Jabotinsky (1880–1940), statesman, author and a founding leader of the Zionistic movement: "Nothing imaginable is impossible; if possible, it could be easily achieved". Jabotinsky did not imagine genetic transformation; moreover, being an outstanding orator, his spoken words sounded more convincing than his written ones, but, with respect to plant genetic transformation, Jabotinsky's spoken words are probably true. Well, up to a certain limit.

2.2. Main Consideration of Plant Genetic Transformation

In the previous chapter, we indicated that all prokaryotic and eukaryotic organisms share the same coding system (although their triplet preference may be different). From this fact, we can deduce that the coding sequences (if necessary, with some changes) of genes from any organism can be transferred to plants and serve there for the production of the respective proteins. On the other hand, plants differ with respect to regulatory sequences, upstream and downstream of the coding sequence, from other organisms. Consequently, transformation cassettes for plants should be constructed in a manner that they contain plant-specific regulatory sequences but the coding sequences can be of any organism.

Plants have a great biotechnological advantage over other eukaryotes such as mammals, due to their totipotency. Functional plants can be regenerated *in vitro* from a wide range of sources, not only from egg cells and embryos. Even cells isolated from mature plant leaves and converted into protoplasts (i.e. cells devoid of cell walls), can be cultured and regenerated into plants, from a wide range of plant species (see Galun, 1981). Plants of many species can also be regenerated from undifferentiated callus tissue that is cultured *in vitro*. In several plant species, "embryogenic" callus can be cultured and be maintained as such, until, by changes in culture conditions, these embryogenic calli will differentiate into plants. We shall not detail the tissue-culture aspects of genetic transformation in plants but will note that there are vast differences in the regeneration capability among plants. Grossly, there are plant families such as Solanaceae where regeneration is easy, while in other families such as Gramineae, the ability to regenerate plants is limited to specific tissues, such as the scutellum of immature embryos of wheat (Gosch-Wackerle *et al.*, 1979). In some cases, where regeneration of plants from de-differentiated callus could not be achieved, there was still the possibility to maintain *in vitro* shoot cultures that could serve for genetic transformation (i.e. by biolistic procedures).

The choice of the transformation methodology is very much dependent on the *in vitro* regeneration capability of the given plant species. Thus, for example, in plants like potato and tobacco which have a good regeneration capability even from wounded tissue, *Agrobacterium*-mediated transformation will be the choice procedure, while in banana, biolistic transformation will be the method of preference. The model plant *Arabidopsis thaliana* is very small and each plant produces many seeds. Thus, in this plant, a special procedure was developed: exposing the whole plant, or its inflorescence, to a suspension of *Agrobacterium tumefaciens* (harboring the required transformation vector). Planting the seeds and harvesting the seeds of the subsequent generation, for analysis of transformation events, will then reveal the transgenic plants (see Clough and Bent, 1998).

In the following, we shall first discuss the main plant genetic transformation methodologies and deal with other entities required for the expression of alien genes in transgenic plants. We shall then handle the selection and verification of transgenic plants after transformation. For additional background information, the readers are referred to the review of Hansen and Wright (1999) and to several reviews in the book edited by Hammond, *et al.* (1999), as well as to our previous book (Galun and Breiman, 1997).

2.3. The Biology of Genetic Transformation Mediated by *Agrobacterium tumefaciens*

In the title of a review on *Agrobacterium*-mediated genetic transformation, Hansen and Chilton (1999) coined this bacterium "*gifted microbe*" (the full title "Lessons in Gene Transfer to Plants by a Gifted Microbe"). Whether or not the term "*gifted*" fits an organism that preys on another organism, in a tricky manner, is a matter of attitude. We take the selfish view of man and accordingly labeled the description of the *Agrobacterium* transformation system as: "*from foe to ally*".

2.3.1. *An abbreviated history*

Readers interested in the history of the *Agrobacterium* system may find it in Galun and Breiman (1997) where many of the original studies were cited. Here, we shall abbreviate it to illustrate the major discoveries that caused the change in attitude towards this bacterium: "*from foe to ally*". The plant disease commonly termed *crown gall* was reported in early times, and recorded by Aristotle and Theophrastus several centuries BC. About 100 years ago, the tumorous out-growth, typically appearing on the upper part of roots, near the stem, was observed in many crop plants (annuals and perennials). The first scientific evidence, based on experiments, was reported two years before the term *genetics* was invented. This report was of Smith and Townsend (1907). It was printed in the journal *Science* 22 days after its submission. It furnished proof that crown gall is caused by a bacterium. For about 30 years, this bacterial disease was handled by plant pathologists until it was picked up by Armin Braun in the early forties in an attempt to use it as a model for human cancer research (see review, Braun, 1969). The biochemical-molecular approach to understand the *Agrobacterium* tumor-inducing system was then moved from New York eastwards across the ocean to the lowlands of Europe (i.e. Belgium, the Netherlands and France), and also to the west (Washington State). In the early eighties, it became clear that this bacterial *"foe"* is useful for the genetic transformation of plants and thus the *"foe"* became an *"ally"*. From publication records, we suspect that in this process of *"from foe to ally"*, some investigators involved in elucidating the system were converted from allies to foes. Here, we refer the readers to the motto of our book (*"The Envy of Scholars....."*).

2.3.2. *The essence of A. tumefaciens biology*

Agrobacterium tumefaciens is a soil bacterium and, as common to bacteria, it may harbor one or more plasmids in its cell. If it contains a tumor-inducing (Ti) plasmid, that is, a circular double-stranded DNA of about 200 kbp, it is potentially pathogenic. We shall mention below that *A. tumefaciens* species are classified according to the type of opines that are synthesized in their host after infection. Opines are modified amino acids that cannot be metabolized by plants. There are slight differences in the structure of the Ti plasmid carried by the different strains of *A. tumefaciens*. Generally, the Ti plasmids have the following components (Fig. 4). First, there are two "border" sequences of ca. 25 nucleotide each at the right border (RB) and the left border (LB) of a fragment of this plasmid. There is a difference between the arrangement of the borders of *octopine-* and *nopaline-* type agrobacteria (octopine and nopaline are two opines). The following description is for the *nopaline* type. The sequences of these borders are arranged in tandem; they are spaced about 23 kbp from one another, and the sequences are nearly homologous in the two borders. This fragment is termed T-DNA. The T-DNA contains a number of genes that are essential for the parasitism of *A. tumefaciens* on plants. These genes are encoding enzymes that synthesize a plant auxin (indoleacetic acid) and a cytokinin. Another part of the T-DNA

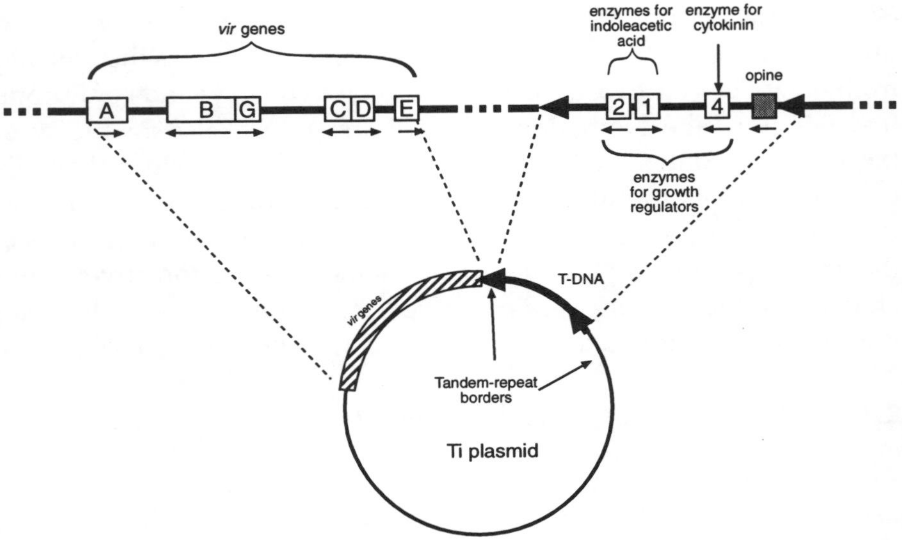

Fig. 4. Genes in the *vir* region and the T-DNA of the Ti plasmid of *Agrobacterium tumefaciens* (based on data from Stachel and Nester, 1986).

codes for enzymes required for the synthesis of one of the opines (e.g. nopaline). The agrobacteria have the capability to utilize the opines. Moreover, there is specificity: the catabolic ability of an agrobacterial species is restricted to this opine for which its T-DNA carries the gene that enables its synthesis in the host plant.

The Ti plasmid harbors, outside of the T-DNA, a number of genes in the *vir* region (Fig. 4) that are essential for the recognition of plant cells, transfer of the T-DNA into the plant cell, as well as its integration into the plant-cell genome, as will be detailed below. Other genes, located on the bacterial genome ("chromosome") also have roles in the parasitizing system, such as synthesizing mediators of bacterial attachment to the plant cell and enhancement of induction of the *vir* genes. These genes are constitutively expressed, while most *vir* genes are induced, to express the respective proteins, only in the vicinity of susceptible and wounded plant cells. Enzymes that catabolize opines are also encoded in bacterial genes and are induced by the opines. The presence of opines also triggers a set of events driven by the bacterial genes in which a *conjugation factor* is an important component, and during which a fibrillar attachment occurs between the bacterium and the plant cell.

2.3.3. *The process of infection*

The aforementioned fibrillar attachment is actually the first step in the *Agrobacterium*/ plant-cell interaction. By a mechanism not yet fully understood but probably mediated by the products of the *att* genes of the bacterial genome, an initial attachment, by means of cellulose fibrils, is established; a net of additional fibrils is then formed that joins additional bacteria to the plant cells. Several additional genes of the bacterial genome are involved in this attachment (e.g. *chv* genes and *psc*A). But the agrobacteria will attach to the plant cell only when the latter have the appropriate binding sites. Here again, the knowledge of the mechanism is not complete but it is plausible that plant cell receptors and their specificity are involved in the host range of agrobacterial species. A scheme of *A. tumefaciens* infection that leads to transformation is provided in Fig. 5.

The wounded cells of susceptible plants release phenolic compounds such as acetosyringone (AS) and β-hydroxyacetosyringone (in tobacco). These phenolic compounds are inducers of the *vir* genes. Actually, the *vir* genes of agrobacteria are induced by these compounds even in the absence of wounded plant cells. But in natural infection, the AS may act as a trigger for chemotaxis, causing the agrobacteria to approach the wounded cells. The bacterium most probably senses the phenolic inducers through the constitutive activity of two *vir* gene products. One protein, VirA, is spanning the inner cell membrane of the bacterium, with its receptor in the periplasm and its kinase domain in the bacterial cytoplasm. When the phenolic compound is attached to the receptor, the kinase phosphorylates the VirG protein, that resides in the bacterial cytoplasm, rendering it

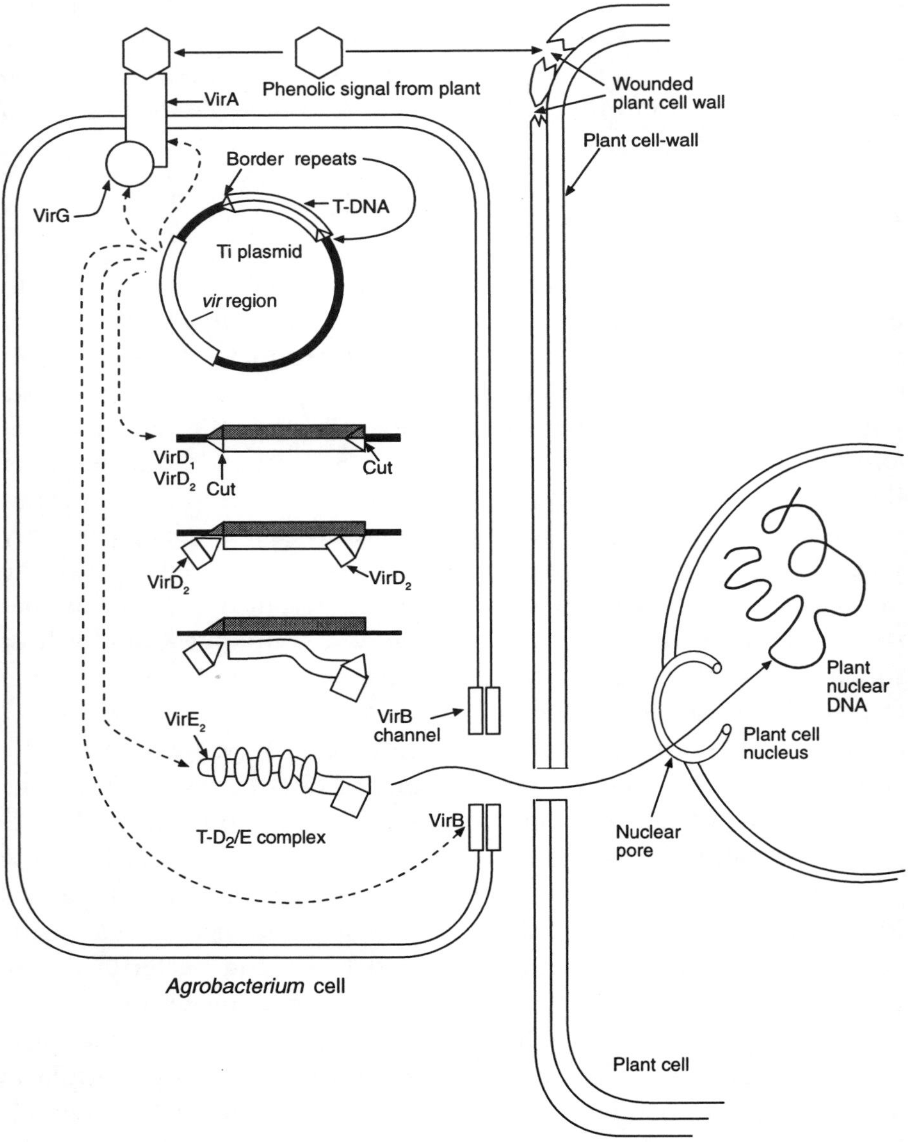

Fig. 5. Scheme of *A. tumefaciens* infection (based on studies of Zambryski, 1997).

active. The VirA protein can probably be further activated by additional effectors. Once active, the VirG protein activates the promoter of several *vir* genes, causing the synthesis of Vir proteins. The latter proteins are active in the processing of the T-DNA, its transfer to the plant cell nucleus and its integration into the host chromosomal DNA. Two VirD proteins are required for the detachment of the T-DNA from the Ti plasmid: $VirD_1$ and $VirD_2$. Of these, $VirD_2$ is active in nicking the "lower" strand of the double-stranded T-DNA region. Nicks are produced in the right and the left border repeats probably between the 3rd and 4th base pairs, counted from the 3′ end of these border sequences. $VirD_2$ binds to the right border, downstream of the nick and keeps its attachment to the single-standed T-DNA. $VirD_2$ then moves with the "lower" single-stranded T-DNA. On the 3′ end, the $VirD_2$ is attached beyond the nick so that it is not further associated with the T-DNA. The "lower" single strand is then detached from the "upper" un-nicked DNA so that a single-strand of T-DNA of about 22 kbp, with the $VirD_2$ at its 5′ end, can now move away from the Ti. A repair replication takes place in the Ti, displacing the T-DNA that was removed. Thereafter, another round of nicking and removal of a single stranded T-DNA, with a $VirD_2$ attached to its 5′ end, can take place. The process can be repeated several times. Thus, the same Ti plasmid can yield several mobile T-DNAs.

It is agreed by the investigators of the *A. tumefaciens* that a single-stranded T-DNA fragment is formed and that this is the infective entity. But whether or not this T-DNA is complexed with the $VirE_2$ protein before it is moved out of the bacterium is under dispute. According to Zambryski (1997), the T-DNA is wrapped by several hundred $VirE_2$ molecules that protect the T-DNA from degradation and keeps its shape as a thin and long strand. The $VirD_2$ attached to the 5′ end of this thin strand may serve as a pilot for its movement out of the bacterium and into the plant cell. Several VirB proteins are assumed to be instrumental in the secretion of the T-complex from the bacterium by forming an appropriate channel through the two (inner and outer) bacterial membranes. It is not clear how the complex enters the plant cell. But once inside the cell, it is probably complexed with cytoplasmic receptors of the plant cell that are instrumental in the translocation of the T-complex through the nuclear pore into the nucleus.

How exactly the T-DNA is integrated into the chromosomal DNA of the host plant is not yet clear, but mapping the T-DNA integrations in plants indicated that insertions occur roughly at random but predominantly in genes. We should recall that under "natural" conditions, the genes on the integrated T-DNA *enslave* the plant cells. To understand this enslavement, let us have another look at the agrobacterial system. As described above, agrobacteria have many features that render them perfect parasites. They trace wounded plant roots by "sniffing" their exudates. By way of chemotaxis, they approach the plant tissue, attach themselves there, divide and start to express sequentially a great number of genes that reside in their Ti plasmid. A defined DNA fragment (T-DNA) is detached from the Ti plasmid and with the help of products encoded in its plasmid genes (*vir* genes), enters the plant nucleus and is integrated by a sophisticated procedure into the plant nuclear genome. The T-DNA carries with it two kinds of genes. One kind codes for plant growth regulators, causing the transformed plant cells to multiply into a large mass of cells (tumor) and the other "forcing" the plant cells to synthesize opines that the plant cell cannot metabolize, but the agrobacteria have the ability to metabolize them. This enables the agrobacteria to multiply vastly and further parasitize the plant roots. How could such an amazingly intricate interaction be established? We should not be surprised. The interaction between cyanobacteria and fungal organisms has established a very ancient symbiosis: primitive lichens (probably several billion years ago). The interaction between the photobiont and the mycobiont in present-day lichens is also rather elaborated. Briefly, the photobiont provides the metabolites and the mycobiont provides structure and environmental stress protection. The boundary between symbiosis and parasitism is not always well defined: is our relation with crop plants a symbiotic or a parasitic one? We can even ask more specifically: are we, by implanting alien genes into plants, enforcing them to produce therapeutic products for our benefit, parasitizing on these plants, or, because we also propagate the transgenic plants, are we having a symbiotic relationship with them? Luckily, by consensus, we have no ethical inhibitions when we prey on photosynthesizing organisms, neither do we have evidence that bacteria have such inhibitions. Turning back to *Agrobacterium tumefaciens,* this bacterium may well be considered *gifted,* probably elaborating its *foe* characteristics for millions of years — but man still has an advantage over the bacterial talents, and in a span of a few years, exploited this *gifted foe* in order to improve his ability to harvest the sun and protect his health.

2.4. The Practice of Genetic Transformation Mediated by *Agrobacterium tumefaciens*

During the early years of studies on the biology of *A. tumefaciens,* these studies were not related to the intention to express alien genes in plants. But in later years, especially during the seventies and early eighties, the idea of expressing such genes in plants was widespread and obviously reached those engaged in *Agrobacterium* studies. In fact, several potential *Trojan horses* were suggested as vehicles of such genes, as, for example, the plant virus cauliflower mosaic virus (CaMV). But the *Agrobacterium* system was found to be the favorite *horse.* Those interested in the sequence of events that led to this development can obtain the relevant information from Galun and Breiman (1997), Zambryski (1997), and Hansen and Chilton (1999). The search for an efficient genetic transformation system was driven by a rather simple aim: to stably integrate a given (alien) gene into the plant genome where it will be expressed. The coding sequence of the alien gene was no hurdle, since, as we indicated above, the genetic code is universal. But this code has to be flanked by regulatory sequences that will ensure the recognition by the transcription and translation mechanisms of plants. Once it was found that the *Agrobacterium* system will insert the T-DNA fragment between the right border (RB) and left border (LB) repeats, irrespective of what sequence is between the RB and the LB — the vehicle was at least potentially at hand. By routine DNA manipulations, the endogenous sequence of the T-DNA (i.e. the genes encoding the enzymes for the synthesis of auxin and cytokinin that cause tumor, and the enzymes for opine synthesis) could be removed and replaced by the alien genes with the appropriate flanking sequences to ensure the regulated expression of the alien genes.

One should recall that when a fragment of DNA, between the RB and the LB is integrated into the plant nuclear genome, it will be maintained during somatic cell division in a heterozygous state. Only upon sexual reproduction may homozygous segregants be obtained and these should then breed true. Ideally, genetic transformation by the *Agrobacterium* system should be straight-forward. One should engineer an alien chimeric gene, with appropriate flanking sequences, between the borders of the T-DNA. This fragment with the borders should be cloned into a plasmid. This plasmid should have a replication capability in *A. tumefaciens* and in *E. coli,* and also two selective genes: one gene for selective replication of the plasmid in the bacteria (this gene should be outside of the T-DNA borders), and one gene for the selection of plant cells that integrated the T-DNA borders; the latter should be within these borders. The replication capability in *E. coli* is essential for the engineering and cloning of the plasmid. The *Agrobacterium* should be pathogenic with respect to containing in its genome or in an additional plasmid all the genes required for the infection and integration of the T-DNA into the plant nuclear genome. Infection with such an *Agrobacterium* line should result in the anticipated transgenic plant. All the above are "ideally" the case, but the reality is very different. In the best hosts (e.g. tobacco leaves, potato tuber discs, tomato cotyledons), only a part of the wounded explants that are infected regenerate shoots in selective medium. In many cases, in spite of selection, the regenerated shoots are not transformed, and then among the transformed and regenerated plants, the expression of the transgene is very variable: only very few of the latter will show a "reasonable" expression of the transgene. Furthermore, even with *A. tumefaciens*-mediated transformation, in addition to the favored single insertion of the transgene into the plant genome, there may be several such insertions. Also, the insertion may not be "clean" with respect to the fact that the borders are flanked by undisturbed genomic sequences. While such "clean" insertions were recorded for the right border, they seem to be rarer for the left border (see Hansen and Chilton, 1999, for discussion and references). There are many explanations for this "reality". We shall handle several of them in the coming sections but we should already indicate here that in order to obtain the transgenic plant of choice, one has to regenerate *many* transformed plants.

For practical purposes, there are many published procedures for *Agrobacterium*-mediated genetic transformation. These methods differ primarily according to the plant species that is used as host. It is beyond the capacity of this book to refer to all of them. But in order to provide the readers with a "feeling" of these methods, we shall provide several examples below. One shall be for tomato, another for *Arabidopis thaliana* and a third for wheat. References for other *Agrobacterium*-mediated transformation methods are provided by recent reviews (e.g. Hansen and Chilton, 1999; Hansen and Wright, 1999). We shall delay the discussion on the regulation of transgene expression, selectable markers, reporter genes and related subjects, to the end of this chapter. These subjects are especially relevant for optimizing the construction of transformation vectors. The type of vector differs according to the transformation methodology (i.e. biolistic versus *Agrobacterium*-mediated transformation).

2.4.1. *Examples of A. tumefaciens-mediated transformation*

2.4.1.1. *Tomato transformation*

The following is a procedure adopted by Dr. Ron Vunsh (Dept. of Plant Sciences, the Weizmann Institute of Science, Rehovot — personal communication). It is useful for several cultivars of tomato such as VF36, Matte and Money Maker.

Tomato seeds are sterilized by exposure for 2 min to 70% ethanol followed by 20 min in 3% NaOCl. Then the seeds are rinsed several times in large volumes of sterile, glass-distilled water. The seeds are then planted in sterile plastic boxes containing solidified Nitsch medium (Nitsch, 1969). When the seedlings attain the emergence of the first true leaf, or before that, they are cut below the cotyledons and the tips are accumulated in a liquid that contains MS (Murashige and Skoog, 1962) minerals and Nitsch vitamins (Jones medium). The 2/3 distal part of the cotyledons is then placed, upside down, on the surface of feeder plates. The feeder plates are 100 × 15 mm plastic petri dishes that contain a layer of solidified (agar) MS medium with 1 mg/l 2.4D and 100 μM acetosyringone, covered with a layer of a logarithmic-phase cell suspension (carrot or tobacco) over which a Whatman #1 filter paper is placed. The plates with the cut cotyledons are maintained for one day at 25°C (low light).

During this incubation period, the *Agrobacterium* cells that harbor the "helper" plasmid with the *vir* genes and the engineered plasmid (containing the modified T-DNA fragment with the transgene and the respective *cis*-regulatory elements as well as the selectable gene) are propagated in 50 ml tubes with 10 ml of 2YT medium containing the appropriate antibiotics. The bacterial suspension is shaken (230 rpm) at 28°C in the dark for one day. At the end of the bacterial incubation, the bacteria are diluted in liquid Jones medium containing 150 μM acetosyringone (pH 5.2). The dilution is done to OD 0.3, and 10 ml of the diluted suspension is put into each of 50 × 10 mm plastic petri dishes. The incubated cotyledons are submerged in this suspension for 15 to 30 min. The cotyledons are then blotted on

Whatman #1 filter paper and placed again on a fresh feeder plate (30 cotyledons per plate), and the plates are maintained either in the dark or under low light intensity for two days at room temperature. Then, the cotyledons are transferred to petri dishes with solidified regeneration medium, composed of agar (0.2%), Jones medium, 400 mg/l carbenicillin (or claforan), 1 mg/l zeatin and appropriate antibiotics (e.g. 100 mg/l kanamycin). The plates are sealed with Saran wrap and maintained in low light at 25°C. Calli, green bumps and shoot initials should appear within two weeks. For further shoot elongation, the cotyledons are transferred to a similar regeneration medium but with less carbenicillin (250 mg/1 or 300 mg/1 claforan). When the shoots have elongated, they are separated from the cotyledons and planted again in the latter type of regeneration medium. A third transfer into a regeneration medium of the latter type, but with reduced zeatin (0.15 mg/l), should result in well-developed tomato shoots. These shoots are transferred into plastic boxes (Magenta) containing 50 ml of rooting medium. The latter is composed of Nitsch medium, 150 mg/l carbenicillin, 50 mg/l indolebutyric acid and antibiotics (e.g. 50 mg/l kanamycin). The shoots will root in about two weeks. Rooted shoots can be gradually transferred into a confined greenhouse. For that, the seedlings are freed of agar, planted first in turf pots in a Saran-covered box. The Saran is removed gradually over five days. Then, the seedlings are planted (with the turf pots intact) in regular pots in the greenhouse. The growing tomato plants (T_0 generation) can then be analyzed (by Southern blot analysis, PCR and/or northern blot analysis), or the plants are grown first until they bear fruits and the seeds are extracted and dried. This T_1 generation can be tested by planting (after seed sterilization) in antibiotic, containing Nitsch medium. Seedlings that germinate within 3–4 weeks should be heterozygous or homozygous for the chimeric transgene that also codes for resistance to antibiotics (e.g. kanamycin). Seedlings that have no resistance to the antibiotics may germinate but are stunted without well-developed roots and usually are highly pigmented with anthocyanin. The latter are discarded and further analysis can be performed with the plants that are developing from the resistant seedlings.

2.4.1.2. *Arabidopsis thaliana transformation*

This procedure was termed *in planta* transformation. While it is described here for *A. thaliana* it can be used for other plants that are relatively small when they attain flowering and that produce many seeds per plant. This procedure differs with many respects from the procedure described for tomato transformation. It was based on findings of Feldmann (1991), developed by Bechtold *et al.* (1993) and then simplified by Clough and Bent (1998). For more details, the latter publication should be consulted. For this method of transformation, *Arabidopsis* plants are prepared by planting about 15 seeds in each of 10 cm pots with soil, that are covered with a screen. After planting, the pots are first maintained for three days in a cold room (4°C) and then moved to a greenhouse with additional illumination to provide long days (e.g. 18-hour light per day). After about one month, rosette leaves are formed and inflorescences should appear. These inflorescences are clipped to regenerate new inflorescences within a week. Then, the plants are ready for transformation. Meanwhile, the *Agrobacterium* suspension is prepared in large quantities (e.g. 2 liters). For that a bacterial suspension is grown in LB with appropriate antibiotics (e.g. 25 mg/l kanamycin). The bacterial suspension is applied after dilution to OD 0.8 and only two components are added to it: sucrose (5%) and a surfactant Silwet L-77 (0.05%) (OSi Specialities, Danbury, CT, USA).

When the inflorescences (after one clipping) are 2–10 cm high and attain a stage at which the youngest flowers are still before anthesis and there are already some fruits, the plants are ready for transformation. The inoculation is performed by turning the pots upside down and submerging the plants in the *Agrobacterium* suspension (with sucrose and Silwet L-77) for 30–60 sec. This dip can be repeated after five days. The plants are then moved back to a confined greenhouse and the mature seeds are collected and planted in a selective medium (e.g. kanamycin-containing Nitsch medium if the vector contained the *npt*II gene for resistance to kanamycin). The rate of transformed seedlings is not high, about 0.5% of the tested ones, but because each *Arabidopsis* plant can produce thousands of seeds and many pots can be transformed in a single experiment — the total number of transformed *Arabidopsis* plants, during one experiment, is satisfactory. This method is especially appealing to molecular geneticists who are reluctant to deal with plant tissue culture procedures.

2.4.1.3. *Wheat transformation*

As shall be described below, earlier transformations of cereals were performed by the biolistic method rather than by *Agrobacterium*-mediated methods. While the former method did yield stable transgenic plants (see Christou, 1996), there are obvious advantages in the latter methods.

In the following, a method that is being developed in our department (Plant Sciences, The Weizmann Institute of Science, Rehovot) shall be described, and is based on our own early studies (Gosch-Wackerle *et al.*, 1979; Perl *et al.*, 1992) as well as on several more recent publications (e.g. Weeks *et al.*, 1993; Nehra *et al.*, 1994; Hansen *et al.*, 1997; Tingay *et al.*, 1997; Wu *et al.*, 1998). The method shall be presented as a guideline rather than as a final protocol since it is still in a phase of modifications towards optimization. The purpose of presenting this method here is to indicate the several changes in the procedure when compared to the method of *Agrobacterium*-mediated transformation described above for tomato.

The source of tissue for transformation is the immature wheat embryo. For the production of such embryos, wheat plants are grown at optimal conditions (i.e. temperatures between 20 and 25°C) in a greenhouse. When anthesis occurs, the spikes are marked and after 10 to 14 days, the spikes are surface-sterilized and the grains are extracted. The embryos (0.6–1.0 mm long) are separated carefully (without wounding) from the endosperms and placed with the scutellum facing up on a petriplate containing solidified callus-inducing medium (e.g. with 2 mg/l 2.4D) for 3 to 5 days. The immature scutellum of cereals excels in its capacity to produce callus that can then be regenerated into plants. An indication that a callus is produced under the upper layer of the scutellum is the appearance of bumps. The scutelli are then transferred for 4 h to the same medium but with higher osmotic value (e.g. containing 0.5 M mannitol). Then, 50 scutelli are placed densely together in a petri dish with the same medium but also containing 100 μM acetosyringone. The dish is put in a particle-shooting device (see Sec. 2.6. Biolistic Transformation) and bombarded with 1 μm diameter gold particles. Then, the scutelli are exposed for 15–30 min to a liquid suspension (with acetosyringone and mannitol) that contains the *A. tumefaciens* cells. These cells should contain one of several helper plasmids as well as T-DNA vector plasmids similar to those used for transformation of dicots (i.e. of tomato cotyledons). There are several super-virulent agrobacterial lines with helper plasmids that are more potent on cereals than the standard commercial lines. Moreover, it is desirable to add *vir* genes (*vir*D_1, *vir*D_2, *vir*E_2 and possibly *vir*G) that will be expressed during the transformation process. Also, the promoter for gene expression in wheat (as well as in other cereals) differs from the promoters used in dicots. The former should include cereal promoters and contain introns of cereal genes. Finally, the typical selectable gene used with dicot transformation, the *nptII,* that causes resistance to kanamycin, is not recommended for cereal transformation. Instead, the *bar* gene that causes resistance to phosphinothricin (PPT) or any other gene for resistance to a compound for which wheat is sensitive, should be included in the vector. After exposure to the agrobacterial suspension (that has a density of about OD 0.5), the scutelli are blotted, maintained for a few more hours in high osmoticum and then transferred to a solidified culture medium (still without a selective agent) for a few (2–3) days. Then, the scutelli are cultured for a period of about two weeks in a solidified medium that contains several components: 2.4D, an antibiotic to kill the agrobacteria (e.g. carbenicillin) and a selectable agent (e.g. PTT, "Basta"). During this culture and subsequent similar culture periods, callus should appear. At a later stage of culture (several weeks), there should be a great difference in callus size among the cultured scutelli. The largest calli are then transferred onto a solidified medium that contains zeatin-riboside as well as the antibacterial agent and the selective agent, but no auxin. This culture should induce shoot production.

The shoots are then separated from the calli and rooted in appropriate solidified medium. The rooted shoots are transferred to turf pots, that are first protected from dehydration, and then transferred to the greenhouse. Wheat seedlings are more sensitive to this transfer than tomato seedlings, thus they are only gradually transferred to standard soil pots in the greenhouse. Once in the greenhouse, leaf samples can serve to detect putative transformed plants by Southern blot hybridization or by the polymerase chain reaction (PCR). Putative transgenic plants are grown to maturity and their grains are harvested. The pollen of such plants can be used to cross-pollinate a standard wheat cultivar. Further testing for the expression of the desired transgene is then performed on seedlings that germinate from the grains of self-pollinated and cross-pollinated plants.

From the above described procedure, it is clear that wheat genetic transformation is much more laborious than the tomato transformation procedure. Moreover, a "yield" of one transgenic wheat plant out of about 1,000 immature embryos is already regarded a success. The mere extraction (by dissection) of 1,000 immature embryos from about 50 spikes requires several hours of intensive work.

2.5. Genetic Transformation Mediated by *Agrobacterium rhizogenes*

The soil bacterium *A. rhizogenes* is a close relative of *A. tumefaciens*. The biology of the former was studied much less than that of *A.*

tumefaciens but it is quite clear that they share common features. *A. rhizogenes* also contains a large plasmid that is termed Ri (rather than Ti in *A. tumefaciens)* and a fragment of this plasmid (T-DNA) is integrated into the nuclear genome of infected plants (Chilton *et al.*, 1982; Tepfer, 1984; Durand-Tardif *et al.*, 1985) where it induces the synthesis of opines and a change in morphology, including the acceleration of root growth. Once transformed, the infected plants transmit these features to their sexual progeny even in the absence of the agrobacteria.

David Tepfer reviewed this system (Tepfer, 1995a) and provided experimental procedures for transformation with *A. rhizogenes* (Tepfer, 1995b). He rightfully claimed that this transformation system is relatively simple and does not require sophisticated laboratory equipment. But he overplayed his claim. He stated that this transformation can be performed "... by using equipment found in everyone's kitchen". However this kitchen should include not only a pressure cooker but also a sterile hood and "a fire extinguisher should be close at hand". Such a "kitchen" may appear too hot for some investigators.

There are two main goals for using *A. rhizogenes* mediated transformation. For one, it is aimed to establish a fast-growing root culture in a plant species that produces pharmaceutically valuable compounds in its roots. The other aim is to express in the roots of transgenic plants an alien gene that will either induce the production of compounds that are not produced in the non-transformed plants or will enhance the production of compounds that are already produced in the non-transformed plants. We shall provide examples of experiments that intended to achieve these goals. Here, we shall note that in the case where it is intended to induce the production of a compound that is not naturally produced in a given plant, the *A. rhizogenes* has to be supplemented with a vector that contains the required transgene within its T-DNA borders. In this case, a selectable gene should also be included within these borders.

The transformation procedure for *A. rhizogenes* is indeed rather simple. It can be performed in either of two ways. In one procedure, the plants are grown in regular pots as seedlings and at an appropriate developmental stage, sterile tooth-picks are first dipped into a suspension of *A. rhizogenes* and are then used to prick the young stems of the plants. After a week or two, callus is produced at the wounded spot and then transformed roots start to grow. These roots can then be used for root culture. Such roots will grow rather fast (e.g. 1 cm per day) even in the absence of growth hormones. If the bacteria harbor a plasmid with engineered T-DNA that contains resistance to an antibiotic compound, the transgenic roots can be selected in a medium containing this antibiotic compound. The agrobacteria can be eliminated by carbenicillin or a similar agent.

The other way to cause transformation by *A. rhizogenes* is to grow seedlings, and then surface-sterilize them and cut sections of the shoots of these seedlings. The sections are submerged in the bacterial suspension (for one or more hours), they are then blotted, rinsed with sterile water, blotted again and planted in solidified culture medium with an antibiotic to eliminate the agrobacteria (and possibly another antibiotic compound for selection). No growth regulating compounds are required in the medium. When roots are formed, they are dissected and recultured. Those that show fast growth and are highly branched are considered transgenic. They are transferred again to solidified culture medium and subjected to analysis to verify that they are transgenic. After verification, the roots can be transferred to liquid culture.

Such root cultures, also termed *hairy roots,* can be maintained for long periods at low temperature (4 to 15°C). Some root cultures resulting from this procedure may spontaneously regenerate shoots so that this could be a way to obtain functional transgenic plants. We do not recommend this procedure as a standard method to produce transgenic plants.

2.6. Biolistic Transformation

Biolistic transformation is a term coined by the inventors of this transformation (Sanford, J.C., Klein, T.M., Wolf, E.D. and Allen, N.K., 1987). The term stands for *biological ballistics*. The idea behind this method was that if one delivers a gene (i.e. a coding sequence with due flanking regions) into a cell, the gene will be expressed in this cell and possibly be integrated into the cell genome, ultimately causing stable transformation. When the transformed cell is then a source of a functional plant, a transgenic plant that may transmit the transgene to its sexual progeny can be secured. By retrospect, this was a wild goose chase but it became a reality. The biolistic system was recently detailed and reviewed by Finer *et al.* (1999) and was also handled by us previously (Galun and Breiman, 1997). Several chapters in the book edited by Potrykus and Spangenberg (1995) are devoted to biolistic transformation. The system can be applied to a wide range of organisms such as algae, fungi, oomycetes and animals but here, we shall restrict our discussion to plants. The principal concept of biolistics is that DNA (or in some cases, RNA) is delivered into target cells, possibly into the plant nuclei, mitochondria or chloroplasts. It may be expressed right after delivery by due transcription and translation (*transient expression*) or it may first be integrated into the plant DNA (in the nucleus, the chondrion or the plastome) and then expressed accordingly. The initial concept of the investigators who invented this method was naive since they did not know how the delivered DNA will be integrated into the plant DNA (in fact, it is not clear up to today). The main advantage of the biolistic approach over previously suggested delivery means (e.g. microinjection, introduction into protoplasts or cells by various chemical manipulations) was that a large quantity of DNA could be targeted into many cells by one shot. For this, the inventors recruited ballistics: coating microprojectiles (tungsten or gold powder of about 1 mm in diameter) with the DNA, and accelerating these microprojectiles to a velocity that will cause them to penetrate one or more cell layers.

Still, the most popular device to perform biolistic transformation is the commercially available helium gun, termed PDS-1000/He, that is marketed by BioRad and licensed by E.I. du Pont de Nemours. The main physical features of this device are that the shooting is triggered by helium pressure that can be controlled and suddenly released to propel a macroprojectile. The macroprojectile carries a "drop" of metal powder impregnated with DNA. After a certain distance, the macroprojectile is stopped but the drop of powder continues until it reaches the target tissue. To maintain the high velocity of the powder, the plant (or plant tissue) is maintained, just before shooting, in reduced air pressure (partial vacuum). Other shooting devices were suggested, either commercial ones or "home-made". Such devices were summarized by Finer *et al.* (1999). For our purpose, we will discuss when the biolistic transformation should be preferred over other transformation methods (e.g. *Agrobacterium*-mediated transformation). For that, we shall consider the main advantages and disadvantages of biolistic transformation as compared to *Agrobacterium*-mediated transformation. The first advantage of the former method that comes to mind is the transformation of the so-called *recalcitrant* plant species. This means plants that investigators were not able to transform with *Agrobacterium,* or that the latter transformation was very difficult. The biolistic system is not dependent on the intricate interactions between the pathogenic bacterium and the plant host. Another advantage of the biolistic transformation is the relative simplicity of the engineering of the transforming DNA: one has to engineer a simple plasmid that contains in addition to a selectible gene, the required coding region, and flank it with regulatory sequences (e.g. promoter and terminator). If this is performed with a commercially available plasmid such as "Blue-Script" KS that contains an origin of replication and a selective gene for multiplication in a standard (*E. coli*) bacterium, the engineering is swift.

On the other hand, the biolistic transformation (as well as other "direct" means of transformation) has a severe drawback. While the T-DNA of agrobacteria is rather precisely integrated (at least at its right border) into the plant DNA, and commonly, only a very few T-DNAs (one or more) are integrated into each plant genome, the integration of DNA during biolistic transformation is unpredictable. For yet unknown reasons, not one but a chain of DNA fragments are frequently integrated into the plant genome. It appears that the DNAs delivered by biolistics rearrange or recombine with each other or yield chains of plasmids and plasmid fragments before being integrated. This may lead to many integrations per genome and consequently result in *co-suppression* (see below) and other complications, rendering this transformation random and unpredictable with respect to the DNA integration events (Hadi *et al.*, 1996). Another drawback of biolistic transformation is the intellectual property problem. The concept of metal powder delivery of DNA is patented and so are the machines that are commercially available. It will take until about 2010 for these intellectual property restrictions to be relieved. True, there are also some commercial restrictions in *Agrobacterium*-mediated transformation (e.g. use of certain vectors). Thus, any endeavor to manufacture commercially valuable products by genetic transformation should take into account the expenses of juristic counseling.

In addition, there were important developments in transformation methodologies that should affect the decision as which transformation methodology is preferable. Let us take one example. In the early nineties, many plant species were considered "recalcitrant" to *Agrobacterium*-mediated transformation (note that the term *recalcitrant* was previously an adjective to describe plant tissue that was difficult to regenerate into whole plants and only recently, it was used to describe plant species that do not yield to genetic transformation). Actually, it was claimed that "most monocots" cannot be transformed by *Agrobacterium.* This was obviously a misleading statement. It was based on reports that species of the Gramineae family are not natural hosts of *Agrobacterium.* It then became evident that all major cereal crops (i.e. of the Gramineae family as maize, rice, barley and wheat) can be transformed by agrobacteria, provided some modifications in the method are applied (see details and references in Hansen and Chilton, 1999). In some specific cases, such as the genetic transformation of chloroplasts, agrobacterial transformation is probably not applicable (Maliga *et al*, 1990). We shall devote a specific section to chloroplast transformation.

The choice may finally be a combination of biolistic and *Agrobacterium*-mediated transformations. We presented above an example of this combination for the genetic transformation of wheat. A somewhat similar *argolistic* approach was proposed for the transformation of maize (Hansen *et al.*, 1997). Table 1 provides comparative information of biolistic and *Agrobacterium*-mediated transformations.

Table 1. The main characteristics of biolistic and *Agrobacterium*–mediated transformations.

	Biolistic	*Agrobacterium*-mediated
Range of applicability	Can be applied to any plant species but it is required that the target tissue can regenerate into functional plants	It is very efficient in some species (especially of the Solanaceae family), less efficient in other species and difficult in some Gramineae crops
Construction of transformation vectors	Relatively simple	Requires cloning into the T-DNA borders of *Agrobacterium* plasmids
Integration into the plant genome	Usually many integrations into each trasnformed genome	Commonly one to a few integrations per transformed genome
Expression of transgenes	May be problematic because of integration into "silent" chromosomal regions and other problems	There are (usually) no expression problems but the level of gene expression may vary considerably between transformed plants

2.7. Transformation of Chloroplasts

Chloroplasts are the main harvesters of light energy from the sun. Although their efficiency is low and a major enzyme active in the reduction of CO_2 (Rubisco) is sluggish in its operation, one should recall that Rubisco is the most abundant protein on earth. Thus, if one can "conceal" a high-value polypeptide, in Rubisco or in any other proteins of the chloroplast, this can lead to a feasible manufacture of medical products. True, inserting alien polypeptides in such a large protein requires the addition of flanking peptides that will be recognized as targets for specific proteases as well as peptides that will serve to retain the valuable polypeptide of the digested protein in a preparative column. Such and other approaches render chloroplasts a potential target for genetic transformation. Attempts to transform chloroplasts by agrobacteria were apparently not effective (see Maliga *et al.*, 1990). But success was reported by means of "direct" transformations: biolistic transformation and introduction of DNA into protoplasts by polyethylene glycol. In the latter method, the protoplasts are then cultured and after selection, they are ultimately regenerated into functional transgenic plants.

Pal Maliga and associates (Svab *et al.*, 1990) pioneered the transformation of chloroplasts in plants by the use of biolistic methods. The events that lead to this transformation may be observed from an unexpected angle. Maliga, then at Szeged, Hungary, pioneered the isolation of a tobacco mutant with resistance to *streptomycin* (in 1973). This mutation was maternally inherited and turned out to be located in the plastome. The mutation was thus very useful as a chloroplast marker in studies by Maliga and associates as well as by one of us. Moreover, the mutation could be traced to a change in a deoxynucleotide in the plastomic gene that codes for the 16S ribosomal RNA (Galili *et al.*, 1989). After many years at Szeged, Maliga immigrated to the USA and became a staff member at the Waksman Institute of the Rutgers University in New Jersey. This institute carries the name of the inventor (or one of the two inventors) of streptomycin: Salman Abraham Waksman. This invention, in 1944, caused quite a stir at the time. It was the first antibiotic compound after penicillin, and had a wide range of activity (against Gram-positive and Gram-negative bacteria). Waksman was suggested for the Nobel Prize (for 1952). But then friends of a PhD student of Waksman interfered, arguing that their friend Albert Schatz was actually the one who isolated the *Streptomyces* strain that produces streptomycin. Albert Schatz contested the patent rights that listed Waksman as sole inventor. Schatz lost with respect to the Nobel Prize (only Waksman received it) but gained part of the royalties by an outside-of-court settlement. But Schatz went south to Chile to get a position. Now, back to Pal Maliga at the Waksman Institute. Maliga made use of the same streptomycin resistance to select cells that, after transformation, integrated the respective coding region. The initial transformation system was performed with tobacco plants that were sensitive to both streptomycin and spectinomycin. Leaves of such plants were bombarded with tungsten powder impregnated with plasmid vectors. The latter included coding sequences for the 16S rRNA that contained the mutated sites causing resistance to either both streptomycin and spectinomycin or only to streptomycin. When the vector with the genes for resistance to both spectinomycin and streptomycin was used and selection of the calli was imposed by spectinomycin in the medium, plants that were homoplastomic for spectinomycin resistance were derived. The antibiotic resistance was maternally transmitted. This transformation approach was investigated further (Staub and Maliga, 1992) and the authors concluded that there is a reasonable level of homologous recombination in the DNA of the plastome so that a sequence of deoxynucleotides flanked by sequences in the plastome can be exchanged. If the region is in the *inverse-repeat* part of the plastome, a "differential copy correction" may cause both repeats to contain the transgenic mutation. Furthermore, they assumed that selection pressure causes sorting out during cell division so that a transgenic homoplastomic state can be obtained. Maliga's team used this transformation method in subsequent studies.

In one of these, they showed that the expression of an important chloroplast protein (32 kDa, D1) is regulated by translation rate rather than by the transcription level of the respective *psb*A gene.

The "direct" transformation of protoplasts by polyethylene glycol treatment (see Koop and Kofer, 1995) was used by the Maliga team (Golds *et al.*, 1993) and by the Hungarian colleagues of Maliga who stayed in Szeged (O'Neill *et al.*, 1993). The latter used the same system. Even the same double resistance to streptomycin and spectinomycin tobacco mutant was utilised. An appropriate fragment of plastome DNA from this mutant was used for the transformation of *Nicotiana plumbaginifolia*. Basically, the same results were obtained and the investigators stated that "... this method for stable chloroplast DNA transformation is comparable with or more efficient than the particle bombardment techniques." Then, came the answer of the Maliga couple (Svab and Maliga, 1993) who reported "a 100-fold increased frequency of plastid transformation..." by using a different selectable gene, *aad*A, that also confers resistance to streptomycin and spectinomycin. The polyethylene-glycol-protoplast transformation procedure to obtain transplastomic plants was developed further by Koop *et al.* (1996). It was recently applied by De Santis-Macioszek *et al.* (1999) to disrupt specifically plastomic RNA polymerase genes. In the latter study, the procedure was further modified. The isolated tobacco protoplasts were first exposed to a PEG solution and to the transforming plasmid. They were then trapped in a grid with solidified alginate and the grids were transferred into liquid culture. The selection procedure was the same as that originally used by Maliga and coworkers (i.e. streptomycin and spectinomycin). The rate of transformation was about 25 transgenic lines per 10^6 cultured protoplasts. The Maliga team extended the chloroplast transformation studies to *Arabidopsis* and rice (e.g. Sikdar *et al.*, 1998; Khan and Maliga, 1999). An update of chloroplast transformation with additional reference on this subject was provided by Bock and Hagemann (2000).

Finally, there is an apparent peculiarity in chloroplast gene expression. While in the expression of nuclear genes, the polyadenylation serves to extend the life-span of the transcript, the opposite appears to occur in the chloroplast. Kudla *et al.* (1996) reported that polyadenylation of chloroplast transcripts prompted their degradation. This phenomenon appears strange at first sight. But we should recall that the chloroplast gene expression machinery has several features in common with prokaryotes and that also in bacteria, polyadenylation can cause degradation of transcripts.

Agrobacterium-mediated chloroplast transformation was reported much earlier by De Block *et al.* (1985). From this report that still requires validation, it is not clear if the work was really aimed to transform chloroplasts or this happened to be the result of one of the transformation experiments. The investigators constructed an *Agrobacterium* transformation vector. They flanked the gene for chloramphenicol acetyltransferase (*cat* gene) by the *Nos* promoter but also retained a bacterial promoter sequence ahead of the coding sequence of the gene. The *cat* gene was integrated into the plastome and the authors assume that the bacterial sequence that was retained upstream of the *cat* code was instrumental in expressing the *cat* gene in the plastome. We are not aware of further successful *Agrobacterium*-mediated transformation of chloroplasts.

2.8. Other Methods of Genetic Transformation

In addition to the two main methods of genetic transformation that were described above, namely, *Agrobacterium*-mediated trasnformation and biolistics, several additional methods were proposed. We shall only briefly mention them as most were listed by us previously (Galun and Breiman, 1997) as well as more recently by Hansen and Wright (1999). Several of these "other" methods are based on "direct" DNA delivery and thus have the same disadvantages as biolistic transformation: random and unpredictable integration of the delivered DNA into the plant genome. These include microinjection of DNA into tissue (e.g. cultured embryos) and the use of silicone-carbide whiskers to introduce DNA into plant cells. In species in which protoplasts can be cultured and finally be regenerated into functional plants, the alien DNA can be introduced into protoplasts by electroporation or by polyethylene glycol treatment. A rather efficient procedure of transformation consists of using specific cell lines (such as the BY-2 of tobacco). In this case, the cells are co-cultured with agrobacteria and following a few weeks of selection, transgenic cell lines are obtained. This method does not result in functional plants, but rather in transgenic cell suspensions because established cell lines usually lose the capability to redifferentiate into functional plants.

Finally, there is a general method of "transformation" that leads to transient expression of alien genes: integration of genes into pathogenic plant viruses. When this integration does not impair the viral infectivity and the virus is spread systemically in the infected plant, the alien gene may be expressed transiently throughout the plant. We shall describe this method when dealing with the production of antibodies in plants.

2.9. Considerations Regarding Transgene Expression

The term *gene expression* is ambiguous. It may mean the rate of production (or retention) of the transcript of a gene, or the rate of production (or retention) of the translation product of the transcript, namely the protein. We put the *retention* in parentheses but this is an important issue because there may be a high rate of transcription while the respective mRNA is not stable so that at any time point, there is only very little of it available for translation. The same holds for the protein. It may be produced at a high rate but is quickly degraded so that very little of it is retained. Since, for practical purposes, we are interested in the harvested protein, what we mean by *expression* is the final level of protein. This consideration brings us immediately to the conclusion that if a transprotein constitutes a given fraction of the total protein of a plant organ, it will be more reasonable to express the transgene in an organ with a high protein content (such as storage protein in legume seeds), rather than expressing it in an organ with low protein content (such as leaves of cereals that attain grain maturity).

As shown in Chap. 1, the route from a coding sequence to protein is rather elaborate. There are two main components in this route: the transcription of the DNA into RNA as well as the maturation of this RNA into the respective mRNA and the translation of the mRNA into protein. For nuclear genes, the two components are also spatially separated. The first occurs in the nucleus while the second component is a process that occurs in the cytosol. In the chloroplast (and in the mitochondrion, which shall not be handled in this book), the two components take place inside this organelle. Thus, when we consider the regulation of transgene expression, we should think about both of the components. In the literature, the respective regulations are termed *transcriptional regulation* and *translational regulation*. In practice, our interference, in the expression level of a given coding sequence, is by flanking this sequence with two DNA regions, upstream and downstream, respectively, of the coding region. The upstream region, that goes from the 3′ towards the 5′, starting from the initiation of translation, will be termed in the following as "promoter", although by strict terminology of molecular biology, the promoter does not contain the *leader* sequence that spans the transcribed but not translated sequence. Likewise, our meaning of *terminator* will include all the sequence downstream of the end of the translated sequence; it may extend beyond the transcribed 3′ end of a gene.

2.9.1. *Promoters*

A comprehensive discussion of promoters was provided by us previously (in Sec. 3.3.1 and Chap. 4 of Galun and Breiman, 1997). Here, we shall not repeat this discussion but rather reiterate the main considerations for choosing promoters in constructing transformation vectors. First, we should recall that plant promoters are considerably different from promoters in bacterial, fungal and animal genes. Thus, plant promoters should be used. Then, there is usually a difference between phylogenetically distant plant species: a promoter from a dicot gene will not be very effective in Gramineae species. There are "minimal promoters" that constitute a part of the CaMV promoter and will cause "constitutive" gene expression. These promoters will induce expression in different organs of the plant and without limitation of the developmental stage (e.g. even in cultured cells). Some promoters induce gene expression only in specific organs or even in specific cells of these organs. Such promoters are useful when expression is required to be limited in fruits, seeds, roots, etc. Certain promoters are activated only by specific chemicals such as plant growth regulators (e.g. auxins, abscisic acid, ethylene) or by physical conditions such as toxic oxygen species and water stress. Using plant growth-responsive promoters is generally not recommended because such regulators will affect other endogenous genes. The promoter may interact with protein products from other genes before it is capable of inducing the expression of its downstream-located gene. It is common that two or more domains in the promoter participate in the regulation of its inducibility. Thus, an effective promoter may span over 1,000 base-pairs. Cassettes for plant transformation that contain standard promoters (e.g. the CaMV promoter or the nopaline promoter) are commercially available. These contain polylinker sequences into which a coding sequence can be inserted as well as selectable genes with their due *cis* regulatory sequences. Additional background for the choice of promoters is provided by Fütterer (1995), Gatz (1997), and Gatz and Lenk (1998).

In a book edited by Reynolds (1999), there are several chapters, each dealing with a specific promoter for use in genetic transformation of plants. Among these are promoters that are based on plant genes that are activated during developmental processes; some of these are activated by plant hormones. The advantage of using such promoters is that the components of signal transduction pathways are already present in plants. By their utilization, the transgene can be expressed at a required stage of development. The other group of promoters are those that are responsive to specific environmental signals, such as nutritional compounds, heat shocks and wounding. The third group of promoters comes from non-plant systems such as bacteria, fungi and animals. The transfer of this group of promoters into plants requires a co-transfer of the appropriate transcription factors into the plant. An example of the latter group is the copper-controllable gene expression in plants (Mett *et al.*, 1993; 1996; Mett and Reynolds, 1999). The basic features of this system are the following (Fig. 6).

In this system, a CaMV promoter activates an *ace* gene to produce ACE. The ACE protein is not active until copper activates it. Once activated, the ACE binds to a metal regulatory element (MRE). After this ACE–MRE binding, the complex activates a downstream coding sequence. The disadvantage of this system is that the level of copper application required to induce the promoter may be too toxic for practical use.

An example of a chemical inducer is the ethanol-inducible promoter (Caddick *et al.*, 1998). This system is derived from the fungus *Aspergillus nidulans.* The system requires the constitutive production of a protein (*alc*R) that binds ethanol. After ethanol binds to *alc*R, the *alc*R activates a promoter (*palc*A) which in *A. nidulans* is upstream of the sequence that encodes alcohol dehydrogenase I. In transgenic plants, both *alc*R and *palc*A are required to activate a gene downstream of *palc*A, by ethanol. The level of ethanol required for the activation of this promoter is low and not toxic. The applied alcohol is metabolized in the plant so that the induction is for a limited time period, but the alcohol application can be repeated. Here, plants have a clear advantage over man, they do not become addicted to this inducer.

If a suitable specific promoter is not readily available, one may consider finding such a promoter by due experimental work. Whether or not to do so depends on the potential commercial value of the product to be harvested from the transgenic plant. Such a search is labor intensive but has a fair chance of success. The principal approach

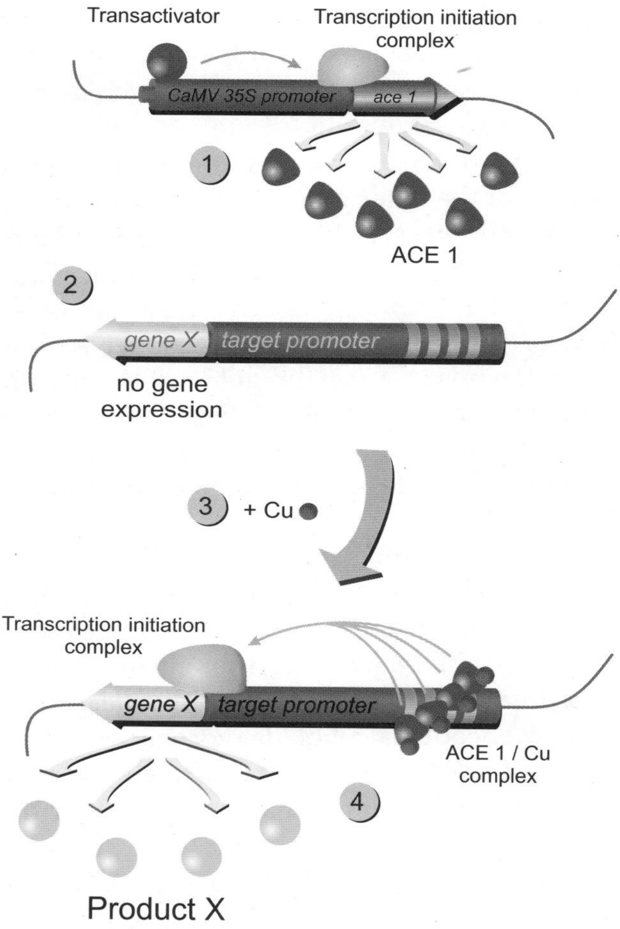

Fig. 6. The copper inducible expression system (1) yeast transcription factor ACE1 is expressed under the control of a strong promoter (e.g. CaMV 35S promoter). (2) In the absence of copper, it cannot bind to its target promoter. (3) The binding of copper induces a conformational change, enabling the ACE1 to bind to specific *cis* sequences of the target promoter. (4) Transcription from the target promoter is induced. For the ethanol-inducible system, ACE1 is replaced by *Alc*R, and Cu is replaced by ethanol. However, it is not clear whether ethanol directly binds to *Alc*R. (From Gatz and Lenk, 1998; reprinted from *Trends in Plant Sciences* **3**: 352–358. Copyright © 1998 with permission from Elsevier Science.)

is *promoter tagging*. This involves the construction of a vector that, in addition to a selectable gene (e.g. *nptII* for kanamycin resistance with the necessary promoter and terminator), also contains a reporter gene (e.g. *gus*, followed by a terminator but without a promoter). This strategy was described by a very elegant research (Barthels *et al.*, 1997) for tagging sequences that regulate nematode feeding structures. The fact that such a study is labor intensive emerges from the 14 investigators (from six laboratories) who collaborated in this study. In this case, the inducers were nematodes but any other induction, biological, chemical or physical, can be used to trace the location where the tagged promoter induces the expression of the reporter gene. Such a study should be performed in a plant such as *Arabidopsis thaliana* where thousands of transformants can be obtained with relatively little effort. Or, because a great number of tagged transgenic *Arabidopsis* lines are available to the public, such lines can be exposed to inductive conditions to reveal expression of the reporter gene at sites of interest and after due induction. Then, the promoter has to be fished out from the sequence upstream of the reporter gene, in suitable transgenic plants.

2.9.2. *Terminators*

The terminators of plants, namely the sequence of deoxynucleotides of nuclear genes that are transcribed but are downstream of the stop codon of the coding region, were detailed by us previously (Galun and Breiman, 1997). We discussed the molecular biology of plant terminators in previous sections of this book, especially in Sec. 1.4.4.3. *The poly-A tail.* We shall therefore only reiterate some main issues about terminators of nuclear genes of plants. Probably, the main point is that our knowledge regarding the specific motifs in the terminators are still meager. We know that terminators contain a signal for polyadenylation (AATAAA) but there may be additional motifs that affect polyadenylation. Another point is that sequences in the terminator may affect the longevity of the mRNA and consequently the level of gene expression. As noted in the previous chapter, the poly(A) and possibly other sequences in the non-translated mRNA, downstream of the coding sequence, are instrumental in the recycling of the mRNA for an additional round of translation. But, contrary to the situation with promoters, where in numerous studies, reporter genes were fused downstream of several promoters (and the promoters in many cases were dissected before fusion), very little information of this kind exists for plant terminators. In one of the few studies on plant terminators, Inglebrecht *et al.* (1996) found a 60-fold difference in the expression of a transgene when different terminators were used. Practically, in most studies in which transgenic plants were produced, the investigators used "standard" terminators such as the nopaline terminator *(Nos)* and the CaMV terminator. These terminators usually fulfil their task but it is feasible that in the future more efficient terminators will become available.

2.9.3. *Selectable marker genes*

The role of selectable marker genes is to select transformed cells after transformation and to eliminate non-transformed shoots, and later seedlings and plants that were not transformed. The strategy of using selectable marker genes differs between *Agrobacterium*-mediated and direct (e.g. biolistic) transformation. In *Agrobacterium*-mediated transformation, the selectable gene is usually integrated, with due promoter and terminator, within the T-DNA borders of the binary vector plasmid. Because the right border is more precisely integrated into the nuclear DNA, in this transformation, the integration of the selectable gene will be closer to the left border while the transgene that is intended to express the required protein is inserted close to the right border. This will reduce the cases in which only the selectable gene will be integrated into the nuclear DNA.

In constrast, for "direct" transformation, the selectable marker genes are engineered into one plasmid while the gene for expressing the required transprotein is engineered in another plasmid. Then "co-transformation" is applied, meaning that, in biolistic transformation, the two plasmids are mixed before bombardment. Experience showed that up to 50% of the plants that are transgenic with respect to the selectable gene are also transgenic with respect to the other transgene.

The most commonly used selectable marker in plant genetic transformation were genes that conferred resistance to antibiotics (such as *npt* II for kanamycin resistance) or to herbicides. A list of these genes is provided in Table 3.1 of Galun and Breiman (1997) where this subject was detailed. Other selectable genes, not based on toxic herbicides and antibiotics, are emerging. One of these is a gene that enables positive selection. By it, only transgenic cells and seedlings that express such a selectable gene will survive. The mannose selection is an example of this type of selectable marker gene (Joersbo *et al.*, 1998). This system is based on the *E. coli* gene (*pmi*) that encodes a phosphomannose isomerase. When this gene is active in transgenic cells and *in vitro* cultured shoots, the latter can be cultured in mannose as the sole organic nutrient. Non-transformed cells and tissues will starve and die.

2.9.4. *Reporter genes*

These genes were very important in the early days of genetic transformation of plants. The expression of these genes is easy to detect by a simple biochemical reaction, staining or even observation after illumination with light of a specific wave length. The coding sequence of the reporter genes is usually derived from non-plant organisms. They are engineered between a plant promoter and a plant terminator. Thus, their expression is a clear indication that the plant is transgenic. They were useful for improving transformation methods and testing *cis*-regulatory elements. A description of reporter genes was provided by the book edited by Potrykus and Spangenberg

(1995) as well as in our previous book (Galun and Breiman, 1997). The first reporter genes were actually the genes for opine synthesis in the T-DNA of *Agrobacterium*. The presence of opines in a transformed plant clearly indicated that the T-DNA was introduced into this plant. Then, the gene coding for chloramphenicol acetyltransferase (CAT) became popular. It required extraction and running the extracted protein on a thin-layer chromatographic sheet. The invention of the GUS (β-glucuronidase) reporter system by Jefferson *et al.* (1986) was a kind of a turning point, because by this method, plant extracts as well as plant tissue could be analyzed. The disadvantage of the GUS reporter is that the analysis kills the plant tissue and the enzyme is very stable. Thus, GUS may be detached in a tissue that is no longer activated by the tested promoter.

Two additional reporter genes became very popular in recent studies with transgenic plants and animals. One is the luciferase gene from fireflies and the other is the *gfp* gene (for green fluorescent protein, GFP) from a jellyfish. In both cases, no extraction or tissue killing is required and the activity of the gene is confined to the cells where the gene is activated. A technical advantage for using these two reporter genes is that they can be readily obtained from commercial companies (e.g. Roche/Boehringer-Mannheim, Clontech, Promega). The commercial products may also have modifications in their coding sequences, rendering them more efficient than the natural genes. The genes are also supplied in plasmids that can be used in the engineering of transformation vectors.

2.9.5. *Other regulatory elements*

There are several regulatory mechanisms, in addition to those noted above (e.g. promoters, terminators and specific factors), that affect gene expression. Some of these are manifested clearly in transgenic plants. These mechanisms are noteworthy; although, as we shall see, in the brief discussion that follows, that in many cases these mechanisms are not well understood and, worse, it may not be possible to plan ahead in order to avoid mechanisms that strongly reduce the expression of transgenes. A comprehensive discussion of these regulatory mechanisms was presented in Chap. 4 of Galun and Breiman (1997). Those interested in detailed discussions on such mechanisms are referred to the following publications: Baumann *et al.*, 1987; Han *et al.*, 1997; Laherty *et al.*, 1998; Bird and Wolffe, 1999; De Neve *et al.*, 1999; Gutierrez *et al.*, 1999; Knoepfler and Eisenman, 1999; Maldonado *et al.*, 1999; Sun and Elgin, 1999; Waterhouse *et al.*, 1999. Here, we shall provide short descriptions of Silencing, SAR/MAR effects and some additional regulatory mechanisms.

2.9.5.1. *Silencing*

Gene silencing in transgenic plants was amply recorded phenomenologically but its molecular basis (or rather bases) is still enigmatic, although some understanding is accumulating. The reviews of Vaucheret *et al.* (1998) and of Wasseneger and Pelisser (1999) discuss this subject and provide ample references. There are two main phenomena of gene silencing termed as *transcriptional gene silencing* (TGS) and *post-transcriptional gene silencing* (PTGS). The former, TGS, occurs when a transgene is integrated into, or close to, a chromosomal region that is "silent", such as the heterochromatin region of chromosomes. It seems that once a gene comes close to a highly methylated region of the chromosome, the methylation is "infectious" — it moves into the transgene and suppresses its transcription. Possibly, the plant can "sense" an alien DNA sequence integrated into its genomic DNA. Thus, silencing can occur when a cereal gene is integrated into a dicot host. Silencing was also observed in *trans*, meaning that an incoming DNA fragment can cause the silencing of an endogenous gene, and possibly vice versa. In practice, it was observed that when several copies of a DNA fragment were integrated into a transgenic plant, the chances of gene silencing increased. Thus, gene silencing will occur more frequently after biolistic transformation than after *Agrobacterium*-mediated transformation. In biolistic transformation, much more integrations per plant genome were commonly observed than in *Agrobacterium*-mediated transformation, and the latter transformation is probably confined more to "active" regions of the chromosome while biolistic transformation causes random integration.

In PTGS, the level of the transcript does not accumulate. It was frequently observed when a strong promoter (35S CaMV) was introduced. It is mainly manifested in T_1 and later sexual generations when the transgene becomes homozygous. Silencing frequently occurred when a transgenic plant was transformed again, especially with a vector harboring the same (strong) promoter or when two different transgenic plants were crossed and the progeny was self-pollinated. This silencing was termed *co-suppression*. While the exact molecular events of co-suppression are not clear yet, there is a lesson: if repeated trasnformations are required, they should not be performed with the same (strong) promoter. But silencing is not always occurring, so if a large number of transgenic plants are secured, some of them will not silence the transgenes. Mutations that are able to counteract the co-suppression (Mittlesten- Scheid *et al.*, 1998) have been revealed. But for practical purposes, such mutations are not yet useful.

2.9.5.2. *SARs and MARs*

The SAR or MAR effects involve chromatin structure and the attachment of the chromatin to nuclear components. The SAR term stands for **S**caffold **A**ssociated **R**egions. The SARs were identified as AT-rich regions. The chromatin in these regions is probably more open for transcription, thus enabling better transcription between SAR domains. They are functionally associated with what was termed *enhancer* sequences. SARs were also termed MARs for **M**atrix **A**ttachment **R**egions. SARs (MARs) may reduce the silencing phenomena. Thus, once these regions become more defined and "universal" MARs become available, they can be synthesized and put on both sides of a transgene in the transformation vectors. Such an approach was tested and it was reported that MARs improved transgene expression (Han *et al.*, 1997). We discussed MARs/SARs previously (Galun and Breiman, 1997) and this subject was updated by Gindullis and Meier (1999).

2.9.5.3. *Additional regulatory mechanisms*

There are several additional regulatory mechanisms of gene expression. Some of these are relevant to the expression of transgenes in transgenic plants (see review of Koziel *et al.*, 1996). Certain introns, especially those spanning into the leader sequence of cereals, seem to have a positive effect on the transcription of transgenes. While the molecular mechanism of such an enhancement of transcription is not clear yet, such introns are already standard components in vectors for cereal genetic trasnformation.

Another regulatory mechanism of gene transcription in plants concerns GT-elements and GT-factors (see review of Zhou, 1999). The GT-elements are usually tandem repeated regions in the promoter that include the consensus core sequence of 5′-G-Pu-(T/A)-A-A-(T/A). The distance between two such elements seem to be critical for the correct binding with a GT-factor. The transcriptional enhancement occurs after correct binding. Such GT-element/GT-factor combinations were revealed in a large number of dicot and cereal species. The binding appears to be species specific meaning there is no good binding of a GT-factor of one species to the GT-element of another species. Thus, at present, the GT-element/GT-factor system is not useful as a general means to improve transgene expression.

2.10. Guidelines for Genetic Transformation

Obtaining reasonable quantities of an alien protein (or oligopepetide) in a specific organ of a transgenic plant is not only a very time-consuming process, it is also rather labor intensive. If the plant species where the transprotein is to be expressed has a sexual cycle of more than six months and one is interested in stabilizing the transgenic progeny up to T_2 or T_3 (i.e. second or third generation after the initial transgenic plant), then the duration of the transformation process is more than a year; possibly two to five years. It is therefore obvious that meticulous planning will pay off.

To demonstrate the type of planning that is required, we suggest the following mental exercise. In this exercise, both *Agrobacterium*-mediated transformation and biolistic transformation will be included. In the former transformation, we shall assume that the binary vector procedure will be applied. In this procedure, the *Agrobacterium* line contains a helper plasmid in which there are genes for the *vir* genes as well as genes for antibiotic resistance activated by bacterial promoters. Then, there is the vector plasmid containing the transgene and a selectable gene within the T-DNA borders. The latter is put into the *Agrobacterium* cells before transformation.

Let us take as example a relatively simple theoretical goal: production of a polypeptide of 100 amino acids that has an antigenic effect in humans and is capable of inducing the production of antibodies against a specific human pathogen. The following considerations then come to mind.

- Where in the plant will such a polypeptide be accumulated in the greatest abundance? One answer could be: in the storage proteins of seeds.
- If indeed seed storage protein is the favored location, there are several further questions. One is which plant species should be used. Another question concerns the ability to integrate a sequence of over 100 amino acids without interfering with the normal accumulation of this storage proteins in the seed. Still, another question comes to mind: what should be the flanking peptides on both sides of the target polypeptide in order to enable its efficient isolation from the total of the seed storage protein. To answer the last question, one should consider sites of proteolysis on both flanking regions and a site for retention in a preparative column that will serve to separate the polypeptide from the bulk of the storage protein.

- Assuming it was decided which plant will serve for the transformation (let it be wheat grains), then the further planning is divided into two separate lines. In one, the methodology of transformation will be considered while in the other line, the vector construction will be planned.
- As for transformation methodology, it may be decided that a combination of biolistic and *Agrobacterium*-mediated transformations will be the procedure of choice; namely a procedure outlined in Sec. 2.4.1.
- The above-mentioned choice requires the availability of an efficient particle-bombardment gun and considerations of the respective intellectual property rights. Also, the choice should be made on the wheat cultivar; again, this requires knowledge of which cultivars can be transformed efficiently and what the intellectual property rights (if any) are of the wheat cultivars that are candidates of the genetic transformation.
- Turning to the construction of the vector, it should be decided which *Agrobacterium* vector strategy to use. Let us assume that the *binary* plasmids is our choice. That means the proper *A. tumefaciens* line (such as a super-virulent line) containing the most effective *helper* plasmid will be chosen. We already know that transformation of Gramineae species is improved by high levels of some Vir proteins (e.g. $VirD_1$, $VirD_2$, $VirE_2$, and VirG). Thus, we should ensure that these are available during transformation.
- The construction of the plasmid with the T-DNA borders is a subject that requires very careful planning. The first decision may be whether to have only one set of these borders in which both the coding sequences for the protein as well as for a selectable gene will be included, or that each of these will be within two different borders. The latter choice requires more elaborate genetic manipulation but has the advantage that the selectable gene can perhaps be separated in future sexual generations from the transprotein.

- For the construction of the fragment that encodes the transprotein, we again have to take into account several considerations: the coding sequence, the promoter (in the broad sense, including the "leader" sequence), trafficking signals, termination sequences and other regulatory sequences.
- The coding sequence should be such that it confirms with the coding choice of the host plant (wheat). This may differ considerably from the coding choice of some human pathogens (e.g. human viruses). Thus, rather than using the respective cDNA of the pathogen, one would prefer that a synthetic DNA is made. The latter DNA should have the proper coding triplets for translation in plants. Furthermore, the synthetic polydeoxynucleotide (of about 300 deoxynucleotides) should also include the codes for the flanking peptides that were mentioned above. The synthetic DNA should also have 5′ and 3′ borders with a restriction site that will facilitate its cloning into the cDNA encoding the storage protein into which this synthetic DNA will be integrated. The synthetic DNA fragment should then be integrated into the plant cDNA that encodes the grain storage protein. In the specific example that we follow, the plant cDNA into which our synthetic DNA is to be engineered already contains the promoter, the terminator as well as the trafficking signals. Thus, we save the effort of engineering them into the vector. But, in other cases, a cereal promoter and leader as well as an appropriate terminator should be engineered upstream and downstream, respectively, of the coding region. Moreover, if the transprotein is to be located in a specific cell organelle or to be secreted from the cells into the intercellular species, the codes for the appropriate signal peptides should be added to the coding region. As a possible safeguard against suppressed expression, one may also consider the addition of MAR sequences.
- Another construction consideration regards the choice of a selectable gene. The choice in many previous cereal transformation experiments was the *bar* gene for the PPT herbicides. One may

tend to avoid the expression of this gene. Thus, there are two choices. One was mentioned above, i.e. to put it within a different set of T-DNA borders and then select, in future sexual generations, plants that do not contain this gene (they should be sensitive to PPT herbicides). Or, to use a different selectable gene, for example, a gene coding for an enzyme that will convert mannose into a sugar that is metabolized by plant cells. Transformed cells will then be selected on a medium in which mannose is the only energy source (see Joersbo *et al.*, 1998).

The above listed considerations should serve only as guidelines and the main purpose of listing them here is to emphasize that one should devote considerable time to planning a genetic transformation in which the aim is to manufacture medical products in transgenic plants. Moreover, our example concerned a specific case where a natural plant protein was to be modified to include the transprotein. In cases where it is intended to add an enzyme activity that may change the endogenic metabolic pathway in a given plant cell organelle or to produce a soluble protein that will accumulate in the cytosol, in the chloroplast or in the mitochondrion, the strategy will be quite different. In these latter cases, more attention should be given to the promoter, the terminator and MAR motifs. Also, if the host is a Gramineae species, cereal promoters rather than standard promoters should be employed.

Chapter 3

Antibodies

3.1. Background on Immunology

A basic knowledge of immunology is required in order to appreciate the research and development involved in the production of antibodies and antigens by transgenic plants. We shall therefore provide a background on the main issues of immunology. Following a very brief historical introduction, we shall deal with the processes of the immune system at the molecular, cellular and organ levels. Immunology has developed a language of its own; we shall thus deal with the immunological terms. We shall see that the immune system has memory. The development of memory, in its widest sense, seems to parallel human development: starting from speech ability that enabled horizontal and vertical transmission of acquired knowledge, through the invention of writing that enabled passage of information among humans and its storage in documents and books, up to the present use of computers and their devices to transmit and store information. But the memory of the adaptive immune system is much more ancient because it is found in birds as well as in mammals and other vertebrates. Thus, it most probably already existed in dinosaurs. Obviously, there is an even more ancient memory, the one stored in the genes and passed vertically from generation to generation. This latter memory is involved with innate immunity. We dealt with this memory in Chap. 1. It is immune memory that we intend to move horizontally from one organism to another by genetic transformation.

Immunology is now a vast field in biomedical science; its scope goes far beyond the short background that can be devoted to it in this book. Therefore, further reading is highly recommended. Several good textbooks are available such as Janeway and Travers (1994) and Roitt *et al.* (1998).

3.2. A Bit of History

The historical origin of immunology brings the *Mother Goose* narratives to mind. Both have to do with milkmaids. In the 18th century, the English physician, Edward Jenner, met a milkmaid. It was during a devastating smallpox epidemic. The milkmaid assured Jenner that she will not be affected by smallpox because she already had cowpox. Cowpox is a mild disease in humans and frequently milkmaids are infected with it by sick cows. The woman told Jenner that it was common knowledge among milkmaids that once you had cowpox, you would not come down with smallpox. Jenner inquired and was convinced that the "story" was true. He then found a volunteer, infected him with cowpox by rubbing a cowpox crust on his skin and the volunteer became immune against smallpox. Jenner termed the procedure vaccination and published the results in 1798. The next great step was about 50 years later, and took place in France (note that the *Mother Goose* narratives were first published in France by Charles Pervault in 1697 and then moved to England and the USA). In the middle of the 19th century, Louis Pasteur was active in various areas of what we now call microbiology. Pasteur's heart was with alcoholic beverages (wine and beer), developing the concepts of pasteurization and microbial fermentation. He was probably the first major contributor to wine production after the biblical Noah. But Pasteur's contribution to immunology was not less impressive than his contribution to fermentation. He found ways to immunize people and animals against several deadly diseases. Back to present times, in 1979, the World Health Organization (WHO) announced that as a result of vaccination, smallpox was eradicated (Fenner *et al.*, 1988). Well, hopefully. So, from Jenner to Fenner, it took two centuries. There were numerous discoveries that followed Pasteur's work and which contributed substantially to the understanding of the immune system and to the application of this knowledge in medicine and animal health care. Moreover, the fields of genetics, microbiology, biochemistry and molecular biology provided insights and tools for the great progress in immunology. Of all the investigators who contributed directly and indirectly to this progress, we shall mention the Russian researcher Metchnikoff who already discovered in the 19th century that microorganisms could be engulfed and destroyed by blood cells termed macrophages. This later mammalian defence was then termed *innate immunity*. Shortly afterwards, there were the studies of von Behring and Kitasato who, in 1890, discovered that in the serum of people who were vaccinated, there are *antibodies* that have the ability to bind specifically to the relevant pathogen. The term *antigen* was then adopted for the entity by which specific antibodies are generated. This was the real beginning of the *acquired* (or *adaptive*) *immunology* studies.

3.3. Innate Immunity

In this and the next chapter, we shall handle the production of antibodies and antigens by transgenic plants. Antibodies are a component of the adaptive immune system. The use of transgenes coding for a group of proteins that belong to the innate immune system will be noted only once as a hypothetical possibility. Therefore, this latter form of immunity will be described very briefly. Though innate immunity is probably an evolutionary early defence against pathogens (it exists in insects that possess no adaptive immunity), it does constitute, even in mammals, the first line of defence and is able to destroy pathogens within minutes to a few days. The immediate action of the innate immunity is manifested among other entities by the presence in the plasma of a set of proteins (complement) that, when activated, are toxic to pathogenic microorganisms. Many such microorganisms and viruses will thus be destroyed. At a further step of the innate immunity, these complement proteins will also coat invading microorganisms. Once coated, the invaders are recognized by phagocytic cells, engulfed and lysed inside these phagocytes (e.g. macrophages, monocytes, neutrophils). The latter process may take a few days. If the destruction by the innate immune system is not complete, then the adaptive immunity goes into action. All human beings are born with the same type of innate immunity and the capacity of this form of immunity (i.e. against which pathogens it is effective) does not change during the life of an individual. The complement proteins are always present in the plasma and their composition is not changed in response to different pathogens. Some bacteria that are resistant to the above actions of the innate immunity may still be destroyed by other acute phase proteins that are produced by the liver in response to infection. Unlike antibodies, the former proteins have a wide-range potency and are not specific to any particular bacteria. Thus, these proteins are also considered as part of the innate immune system.

3.4. Adaptive Immunity

The adaptive immune system is based on two main blood cell types (B- and T-lymphocytes), as well as on other blood cells and several organs and tissues that collaborate in an intricate process. We shall first look at the cells, organs and tissues and then come back to a short description of the process.

3.4.1. *Cell types derived from the pluripotent hematopoietic stem cells*

The principal cell types of the immune system are derived from the *pluripotent hematopoietic stem cells*. A general scheme of the descendants of all these cell types is shown in Fig. 7. The stem cells are also the source of red blood cells. They first differentiate into erythroblasts and then further differentiate into erythrocytes which, in mammals, lose their nuclei in the course of differentiation. The other two cell progenitors derived from the (pluripotent hematopoietic) stem cells are the common *lymphoid progenitor cells* and the *myeloid progenitor cells*. The latter produce the *megakaryocytes* which produce the *platelets* and several types of "white" blood cells that are accessories to the immune system. There are three types of polymorphonuclear leukocytes that are derived from the myeloid progenitor cells: *basophils, eosinophils* and *neutrophils*. The myeloid progenitor cells also produce the *monocytes* that are the source of *macrophages*, the *mast cells* and the natural killer cells. The cells that are of prime importance to the adaptive immune system are the *B-* and *T-lymphocytes* that are derived by different differentiation pathways from the common lymphoid

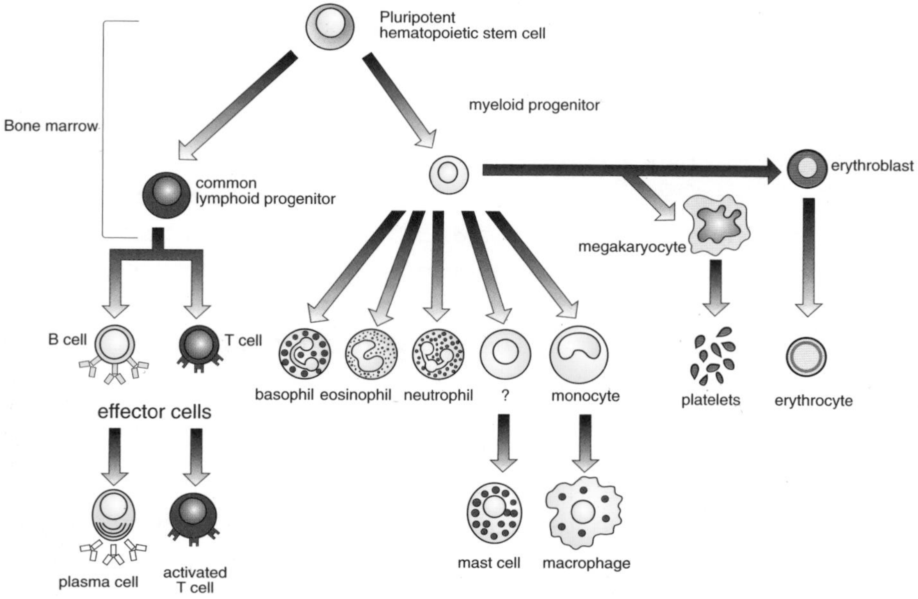

Fig. 7. Scheme of the main cell types descended from the hematopoietic stem cells.

progenitor cells of the bone marrow. These lymphocytes deserve a more detailed description.

3.4.1.1. *The B-lymphocytes*

The B-lymphocytes (B-cells) as well as T-lymphocytes (T-cells) mature in the bone marrow. These cells are the pillars of the adaptive immune system. The B-cells can be defined by the rude German expression as *"Fach Idioten"*: they are single-minded, have a long-lasting memory and are able to fulfil only a limited range of tasks, but these tasks are performed very effectively. Their prime task is to produce one kind of immunoglobulins (antibodies) and expose them on their outer surface as receptors. Following activation, B-cells produce large numbers of these immunoglobulins (antibodies) and secrete them out of the cell. The basic structure of these immunoglobulins is shown in Fig. 8. They can be viewed as Y shaped. Two heavy chains form the Y and two light chains are bound to the arms of the Y, in the heavy chains by disulfide bonds and other bonds. The heavy chain characterizes the class of the immunoglobulin, meaning that the immunoglobulins marked as *IgG* have gamma (γ) heavy chains and those marked as *IgA* have alpha (α) heavy chains. There are also two types of light chains: kappa and lambda. The heavy and the light chains have each a variable region and one or more constant regions as shown schematically in Fig. 8. Since immunoglobulins can be cleaved by partial enzymatic digestion, there is a further division. Digestion with papain (in the presence of cysteine) will result in two separated arms and a stem. The two arms are termed F_{ab} fragments (for **f**ragment **a**ntigen **b**inding) and the stem is the F_c fragment (for fragment **c**rystallizable). Each F_{ab} fragment, which is composed of the light chain and part of the heavy chain, has a unique *antigen recognition capability*. This is due to its variable region. So, the receptors on each B-cell have two identical recognition sites. The peculiarity of the B-cells is that the human body produces a very large number of different B-cells and the difference is in the recognition specificity of the receptors. Thus, the total population of B-cells of a mammal can

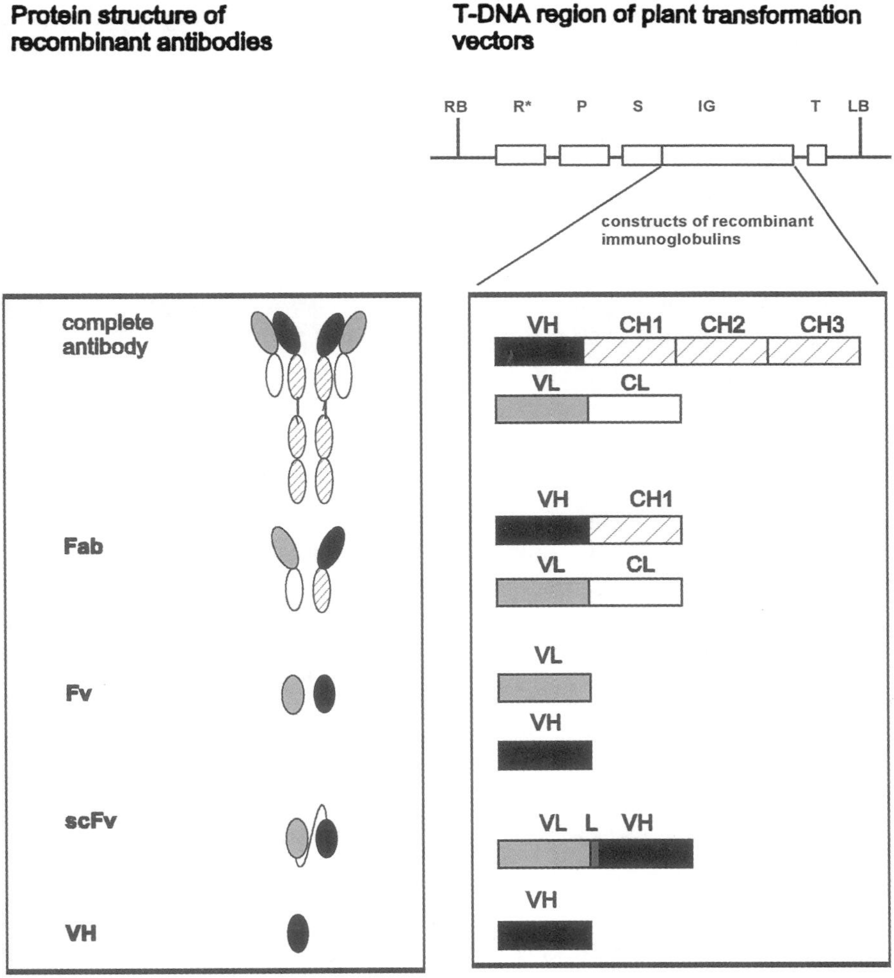

Fig. 8. The structure of antibody and antibody fragments. VH, variable heavy chain; CH1, CH2, CH3, constant heavy chains; VL, variable light chain; CL, constant light chain; L, linker (courtesy of Dr. Rainer Fischer, Aachen, Germany).

recognize almost all kinds of antigens, but each individual B-cell will recognize only one specific antigen. The exact genetic mechanism that is responsible for this situation is beyond our scope but can be summarized as follows. The variable regions of the polypeptides that comprise the light and heavy chains of the immunoglobulins are encoded by hundreds of different genes. During maturation of a B-cell, these segments are rearranged so that the final DNA in this region, that encodes the variable regions of the chains, can have about 10^6 different sequences. The specificity of a receptor is the result of the interactions of the two (light and heavy) chains. Therefore, the final variation of the specificity becomes very large. The DNA rearrangements occur during a limited time of the maturation of the B-cells in the bone marrow. Once the B-cells are mature, there is no further change in the sequence of their DNA. Then comes a period of recognition of *self-antigens* by the receptors. The B-cells that recognize and bind self-antigens are destroyed. The B-cells that did not bind (self) antigens are now ready for policing the body for alien antigens. They are the small resting B-lymphocytes. Once they encounter antigens that are specifically recognized by their receptors, they become activated (*lymphoblasts*). We shall come back later to the process of activation but here, we note that each lymphoblast starts to divide repeatedly so that 1,000 or more cells of the same specificity will become available for binding within a few days. Each of these will produce large numbers of antibodies similar to the original antigen-bound receptors, and secretes these antibodies.

While the variable regions in the arms of the Y-shaped antibodies have the task of recognizing antigens, the *constant* regions characterize the type of antibodies that are produced. Thus, there are five *isotypes* or *classes* of antibodies. Each of these classes has typical sites of action (e.g. IgM acts in the intravascular environment and IgA in the luminar secretions and breast milk). The different isotypes also differ in their functions. Some will agglutinate bacterial toxins, others will attach to bacteria making the pathogens ready for engulfment and digestion by macrophages. There are also differences in the general structure of the antibodies. The scheme in Fig. 8 represents IgG antibodies but, for example, the circulating IgM antibodies consist of five IgM monomers arranged in a pentameric "star" connected at the stems by a J (joining) chain.

3.4.1.2. *The T-lymphocytes*

The origin of T-lymphocytes (T-cells) is also in the bone marrow. But then their destinies and roles are very different from those of B-cells. T-cells migrate to the thymus where they undergo their entire differentiation and where they proliferate and are selected. The mature selected T-cells are then released for circulation. The mature T-cells produce and expose their specific receptors and *co-receptors*. There are two types of T-cells according to their co-receptors (Fig. 9). The receptors and co-receptors of T-cells interact with cell-surface molecules of body cells, which bind an alien peptide (a peptide antigen). These molecules are termed MHC for **m**ajor **h**istocompatibility **c**omplex. There are two kinds of MHC complexes and, respectively, two types of T-cells that bind these complexes. One binds to MHC class I and the other to MHC class II. The establishment of this binding requires the participation of the respective co-receptor of the T-cells. The co-receptor CD_8 interacts with MHC class I and the co-receptor CD_4 interacts with MHC class II. The MHC class I molecules are exposed on all nucleated body cells. Once a body cell is invaded by a pathogen (particularly a virus), it will display pathogen peptides on its MHC class I receptor. When this happens the respective T-cell with its CD_8 co-receptor will bind to this MHC class I and kills the cell. It appears as a cruel system: once invaded by a virus, this body cell is destined for destruction. But it is an efficient way to save the uninfected body cells from the invading pathogens. Cells of the immune system have MHC class II on their surfaces and these are potential targets for the T-cells that have CD_4 co-receptors.

The receptors of the T-cells resemble B-cell antibodies; meaning they are similar to one arm of the Y-shaped antibody. These receptors are heterodimers and have two parallel chains, with variable regions pointing away from the T-cell surface and constant regions close to the cell surface (and a transmembrane region as part of the constant region). The interaction between the two variable regions determines the specificity of recognition. The two chains are connected (in a similar manner as in antibodies) by a disulfide bond at a region termed *hinge*.

3.4.2. *The playgrounds of the adaptive immune system*

The immune system is structured in a way that it would reach virtually all the living cells of the body. But, there are organs and routes that have vital roles in this system.

The bone marrow is the site where the stem cells reside. These pluripotent hematopoietic cells are the progenitors of the red blood cells as well as all the cells of the immune system, as shown in Fig. 7. The bone marrow is also the site where the common lymphoid progenitor cells differentiate and mature into small lymphoid B-cells. The immature T-cells are carried by the bloodstream to the thymus where they mature into the two respective types of T-cells: the killer T-cells (with the CD_8 co-receptor) and the activating T-cells (with the CD_4 co-receptor). The bone marrow and the thymus are also the sites where the respective B- and T-cells that bind self-antigens are eliminated before they migrate to the periphery. The B- and T-cells will meet additional self-antigens outside the bone marrow and the thymus. Tolerance to these additional proteins is achieved in a different way from that in the bone marrow and the thymus.

The *lymphatic system* consists of *lymphatic tracts* and *lymph nodes.* The lymphatic tracts (vessels) lead into the lymph nodes. Antigens as those resulting from pathogen infection are engulfed by phagocytic cells. The latter move along the lymphatics (vessels) into the lymph nodes where they present digested fragments (on MHC molecules) to the lymphocytes (B-cells and T-cells). The naive lymphocytes enter the lymph nodes from the blood vessels. They then exit from the nodes by the efferent lymphatic vessel and enter the blood system

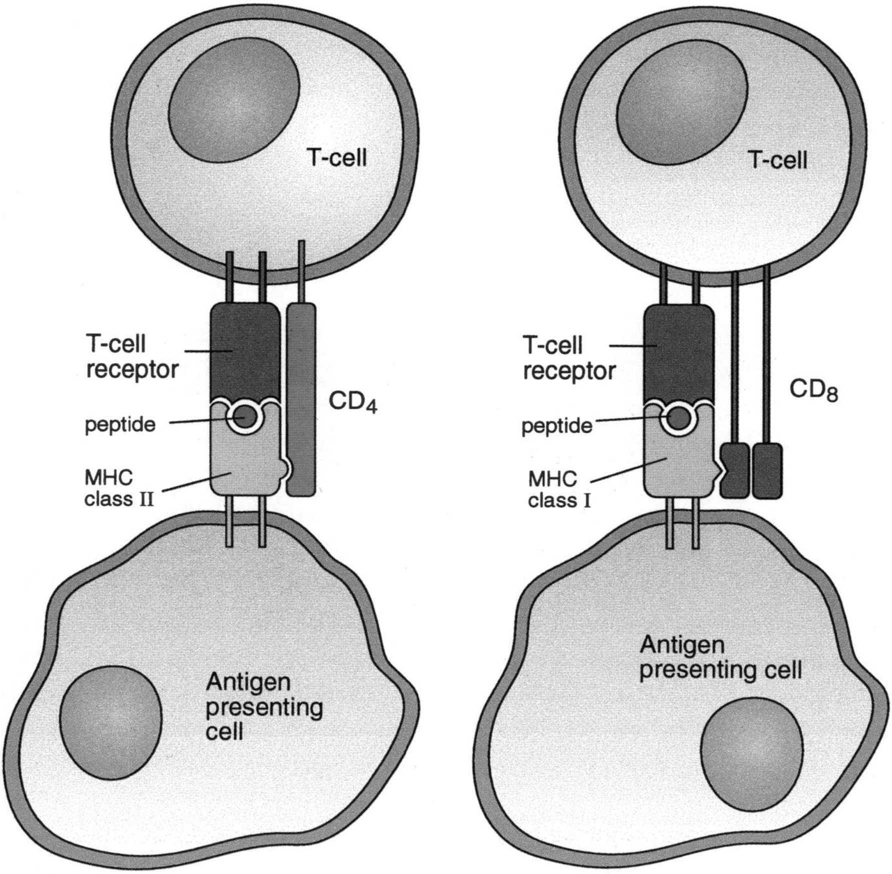

Fig. 9. The receptors and co-receptors of T-cells and their attachment to the respective MHCs.

via the *thoracic duct*. Many thousands of such lymph nodes are located in various places throughout the body so that many "meeting points" between lymphocytes and antigens are available. In addition, there are the spleen, the gut-associated lymphoid tissues, the Peyer's patches and the appendix, where dense populations of lymphocytes will meet antigens. The lymphocytes will consequently be activated and respond accordingly to initiate the adaptive immune response. Lymphocytes that did not meet their specific antigens stay as naive lymphocytes and continue to recirculate from the peripheral lymphoid tissues (e.g. lymph nodes) via the lymphatics into the blood system, and back into the peripheral lymphoid tissues.

3.4.3. *Interactions in which cells of the immune system are involved*

When the receptors on a B-cell bind an antigen, this constitutes the first signal for starting the activation process (i.e. becoming a lymphoblast). There is also a need for activation of the B-cells by lymphokines of the T-cells. After these two signals, proliferation and mass production ensue. The antigen signal can be delivered to the T-cells by either of three antigen-presenting cell types: B-cells, macrophages and dendritic cells. These cells are termed, as a group, *professional antigen-presenting cells,* because they "present" the antigen to the receptors on the T-cell. While the B-cells and macrophages have additional tasks in the immune system, the dendritic cells' role is to present the antigen (mostly after viral infection) to the T-cells. A lymphocyte receiving only the first signal will not differentiate into an effector (or killer) cell. Indeed, receptor binding without the second signal will inactivate the lymphocyte. This is believed to be the basis of peripheral tolerance to self-antigens. There is a main difference between the recognition of antigens by B-cells and T-cells. B-cells "recognize" antigens in a "direct" manner, meaning that extracellular antigens are bound by their surface immunoglobulins. T-cells, in contrast, detect antigens generated *inside* other cells, but the latter present them to the T-cell immunoglobulins on a special "dish" (MHC).

A special type of T-cell interacts with B-cells (as briefly mentioned above). The process starts when the receptors (immunoglobulins) on the surface of a B-cell bind a specific antigen. The receptors with their bound antigen are then internalized and fragmented. Fragments of the antigen are then attached to the MHC class II and returned to the surface of the B-cell. At this stage, the helper (CD_4) T-cells go into action. These T-cells recognize the antigen presented by the MHC class II and their receptors and the CD_4 molecule binds to the MHC–antigen complex. This binding activates the B-cell to proliferate and differentiate into an antibody-producing *plasma cell*.

Pathogens and their toxic products may reside extracellularly or inside the body cells. When they are outside, they reside in the extracellular spaces (or in the blood plasma). Extracellular pathogens can be bacteria, protozoa parasites, fungal organisms or parasitic worms. Such external pathogens (and their products) can be handled by the immune system in either of three ways. In all of these cases, antibodies, produced by B-cells, are involved. The antibodies can bind toxins and by that neutralize them, and the neutralized complex is then phagocytized and ingested by macrophages and polymorphonuclear leucocytes. Note that the macrophages belong to the innate immune system and thus will swallow and ingest any complex that contains antibodies. Antibodies may also attach to the surface of a pathogenic microorganism and such an antibody-covered microorganism can likewise be engulfed and digested by macrophages. The third way occurs when bacteria are in the blood plasma. Then, antibodies (e.g. of the IgM or IgA class) may also cover them, and with the additional effect of the complement in the blood, the bacteria may be lysed and digested and their fragments phagocytized.

The immune system can combat also intracellular pathogens. For example, the bacterium *Mycobacterium tuberculosis* (causing tuberculosis) grows inside macrophages. When such an invasion occurs, the macrophage exposes an antigen on its surface. These antigens are recognized by *inflammatory* T-cells, and the latter will be attached to the infected macrophage. This attachment will signal the macrophage to operate its digestive capacity that resides in its lysosomes. Another example is a virus-infected body cell. The intracellular viruses cause exposure of specific antigens. When a *cytotoxic* (killer) T-cell recognizes such cells, it kills the virus-infected cell.

3.5. Production of Antibodies in Experimental Animals

Because in a subsequent part of this chapter we shall deal with antibodies produced by transgenic plants, we shall first review the production of antibodies in the conventional manner, based on the adaptive immune system of animals.

Antibodies are useful for two main purposes. The first use is for the application of antibodies to patients or sick animals to assist in combating a specific pathogen. The other use of antibodies is as research tools.

3.5.1. *Polyclonal antibodies*

In brief, the conventional means to produce antibodies is to apply the suitable antigen to animals (e.g. rabbits, goats), drain blood from these animals, and separate the serum from the blood cells. The serum is then the source of antibodies. The actual procedure is rather elaborate. First, there is the question of where and how to introduce the antigen to the animal. The application site will determine the type of antibody that will be harvested. Then, for effective immunization, the antigen by itself will usually not be sufficient. If the antigen is a small and simple organic molecule (a hapten), it may not be *immunogenic* , meaning it will not induce the production of antibodies. For induction of an immunogenic response, a carrier protein may be required. Then, the efficiency of immunization may be vastly increased by *adjuvants.* There are several types of adjuvants and they commonly contain water-in-oil emulsions or aluminium oxide gels with additives derived from certain bacteria. Because we are dealing with products of medical and veterinary importance in this book, we shall be concerned with antibodies that bind to pathogens and to pathogenic toxins; these antibodies are elicited by protein antigens. We have no examples yet of nucleic-acid-binding antibodies that are produced by transgenic plants. Only protein antigens can elicit fully developed adaptive immune responses because only such antigens can engage T-cells in the immune system and establish the required immune memory. Protein antigens or haptens with protein "carriers" will lead to a situation that following the first immunization, repeated application of the antigen will further increase the yield of antibodies. This repeated *boosting* by antigen may lead to *hyperimmunization*. The addition of adjuvants is important in the first application of the antigen.

Another issue of eliciting antibody production is the dose of antigen. Commonly, the increase in dose increases the yield of antibodies, up to a plateau level. Then, higher doses will reduce the yield.

The route of antigen application is also a major factor in antibody yield. The route should be taken into account according to the type of antigen used. Commonly, subcutaneous application will be more effective than intravenous application, but in some cases (e.g. poliomyelitis antigens), oral application that will reach the gut-associated lymphoid tissues or even application to the mucosal-associated lymphoid tissue may have an advantage. The serum that is obtained from blood drained from the immunized animal will usually not only contain the antibodies of choice. Therefore, methods of purifying the antibodies are required. Moreover, the receptors on the lymphatic cells bind only a certain determinant (epitope) on the immunogenic protein. Each of these cells will be multiplied into a clone. Since different B-cells will recognize different epitopes that may be exposed on one pathogen, cells of several clones may become engaged in producing (different) antibodies to the same trigger. The result will be *antisera* with *polyclonal antibodies*.

3.5.2. *Monoclonal antibodies*

The concept of producing *monoclonal antibodies* was developed in order to yield a uniform population of antibodies that are derived from a single lymphatic cell. The concept could be perceived after methods to culture mammalian hybrid cells became available. We shall outline the method of producing monoclonal antibodies but skip the technical details. For such details, the readers are referred to the respective literature (e.g. Harlow and Lane, 1988; Ausubel *et al.*, 1999). The production of monoclonal antibodies was conceived by Kohler and Milstein (1975). The principal idea was to fuse two murine (mouse) cells; one cell that produces a specific type of antibody and the other cell furnishing the ability of perpetual division. The first is a spleen cell, derived from an immunized mouse; the other is a myeloma cell ("immortal" cancer cell). After fusion, the fusants (hybridomas) are cultured (on a layer of feeder cells) and the resulting colonies are screened for antibody production. Selected colonies are dispersed and cultured further to yield colonies derived from single hybrid cells. The latter colonies are analyzed again and the ones that produce the desired antibodies are further expanded to yield large quantities of the antibodies. During this procedure, samples of the cultured cells are frozen (in liquid nitrogen). These samples can be recovered when needed for the production of more monoclonal antibodies.

While monoclonal antibodies are mostly used as tools in molecular biology as well as in biochemical and biomedical studies, they may become useful for therapeutic purposes. For example, in patients with small cell lymphoma, the cancerous cells have very specific epitopes on their surfaces. Moreover, each individual patient will have unique antigens. Treatment of patients with monoclonal antibodies that specifically recognize the unique epitopes on the lymphoma cells could specifically destroy these cells without damage to normal lymphatic cells. Here, *time* is a major consideration. It is required that with the first diagnosis of lymphoma, the long procedure of monoclonal antibody production will start. Could transgenic plants manufacture great quantities of such monoclonal antibodies in less time than by the conventional procedure? We shall come back to this question in a later section of this chapter.

3.6. An Epilogue on the Adaptive Immune System

We saw that the ability of the mammalian adaptive immune system to defend the organism against pathogens is based on the Latin maxim: *"Si vic pacem, para bellum"* (this maxim, which roughly means: "If you seek peace, be ready for war", was inscribed over the entrance of the Austrian–Hungarian Ministry of War before World War I, but the inscription was probably not implemented).

The immune system produces an almost unlimited variety of lymphocytes, each having a different antigen receptor. So, the organism is prepared for a vast range of invaders. This huge number of different receptors was enigmatic until the molecular genetics of the genes involved became understood. It emerged that the rearrangements of gene segments by a mechanism (VDJ) similar to the transposable element phenomenon takes place during a limited period of lymphocyte maturation (Roth and Craig, 1998). Immunologists have to be grateful to maize grain-spots for the understanding of this genetic peculiarity. But, they have to be even more grateful to Barbara McClintock who studied these spots by genetic and cytogenetic methods, and developed the transposable element concept (she received the Israeli Wolf Prize and subsequently the Nobel Prize for her discoveries). Interestingly, McClintock's studies, in the early forties, were either ignored or met with skepticism by most contemporary geneticists as her conclusions did not comply with the dogmatic attitude that genes are stable and confined to specific loci on the chromosomes. Only a handful of her previous "Cornell Corn-Breeders" colleagues, who appreciated McClintock's wisdom and integrity, accepted her conclusions. Now, we know much more on the reality of transposable elements and are aware of their widespread occurrence. Recent findings highlight the irony of nature: the same VDJ DNA recombination that is fundamental for human (and other mammals) protection against pathogens may also be the mechanism that underlies various chromosomal aberrations that lead to lymphoid neoplasm (Roth and Craig, 1998).

3.7. Production of Antibody Fragments by Transgenic Plants

3.7.1. *The concept and the biochemical approach*

As indicated in a previous section, antibodies are bivalent molecules composed of two identical heavy chains (molecular weight 50,000) and two identical light chains (molecular weight 22,000). Porter (1959) fragmented an antibody by crystalline papain into two monovalent (F_{ab}) fragments (molecular weight 50,000) and one F_c fragment that is devoid of antigen binding. A further attempt to reduce the size of the F_{ab} was performed by D. Givol and associates. These latter investigators obtained a fragment that had half the size of F_{ab} but retained the same binding capability to an antigen as the regular full-size antibody (e.g. Inbar *et al.*, 1972; Hochman *et al.*, 1976 and see Givol, 1991). These authors chose the mouse myeloma protein-315 that has an IgA which possesses anti-2,4-dinitrophenyl (DNP) binding activity. The variable domains of the two chains (V_L and V_H) of this immunoglobulin were cut away from their respective constant domains by enzymatic cleavage, followed by separation and purification. The fragment obtained was termed F_v. The V_L and V_H were kept together by their noncovalent bonds. Each of the two chains consisted of about 110 amino acids, and the total molecular weight was about 25,000. The F_v indeed retained the binding capability of protein-315. The F_v has obvious advantages over the full-size antibody. The former is much smaller and lacks the constant portion of F_{ab}. It should therefore improve accessibility and avoid interference with the immune system. The schemes of antibody and antibody fragments are presented in Fig. 8.

3.7.2. *Production of F_v in bacteria*

Several years lapsed before the use of F_v became widespread. The elaborate enzymatic cleavage and purification by Givol and associates (which also required a considerable amount of antibody to start the procedure) were replaced by methods that were developed from molecular genetics. In several initial studies, the amino acid sequences of the V_H and the V_L were computer-translated into DNA sequences. These sequences were further changed to provide triplet codons that are preferred by the bacterial protein synthesis system. The sequences were expressed in bacteria and yielded F_v. The respective DNA sequences were then synthesized. Then, a *linker* was devised to strengthen the association between V_H and V_L (or vice versa) by using three repeats of Gly–Gly–Gly–Gly–Ser. A DNA sequence that encoded the linker was then ligated to the two DNA sequences encoding V_L and V_H, respectively, and a single DNA sequence was established. The latter was cloned into an appropriate bacterial plasmid and bacteria were then transformed with this plasmid. This resulted in the bacteria expressing the required *single-chain F_v (scF_v)* (e.g. Bird *et al.*, 1988; Huston *et al.*, 1988). A similar construction strategy enabled the production of a chimeric scF_v. The latter scF_v had a mouse V_H and a man V_L (Jones *et al.*, 1986).

With the advent of the polymerase chain reaction (PCR), the procedure to construct a single DNA sequence encoding the whole scF_v was further simplified (see Plückthun, 1991; Kaluza *et al.*, 1992). A publication with the presumptuous title: "*Man-Made Antibodies*" by Winter and Milstein (1991) (the second author was the co-inventor of monoclonal antibodies) reviewed this subject. This title was perhaps an exaggeration. First, man makes antibodies constantly in his own body. Second, man is assisted by bacteria and mammalian cells in the production of antibodies, as was reviewed in this publication.

3.7.3. *Production of scF_v in plants*

Before scF_v production in plants was aimed to serve therapeutic purposes, such scF_vs were intended to modify plant physiology. Owen *et al.* (1992) used a monoclonal hybridoma cell line that secretes an IgG1 monoclonal antibody which binds specifically to phytochrome, a plant regulatory photoreceptor. An scF_v gene was produced by PCR from the DNAs that encode the variable domains (V_L and V_H)

and a linker (42 base-pairs) that was ligated to the 3′ of the V_L coding sequence. The constructed chimeric scF_v gene was first expressed in *E. coli*. The bacteria accumulated scF_v in insoluble cytoplasmic inclusion bodies from which functional scF_v could be isolated. These scF_v fragments were bound specifically to phytochrome. The authors then moved the scF_v gene into an *Agrobacterium* plasmid, behind the 35S CaMV promoter. The plasmid also had a gene for kanamycin resistance. The plasmid was introduced into the LBA 4404 *A. tumefaciens* strain and the latter was used to inoculate tobacco leaf sections. After tissue culture, 119 putative transgenic plants were obtained. The investigators focused on one such plant that had maximal expression of scF_v. This scF_v had the expected size and its functionality as binder of phytochrome was ascertained by several methods. The scF_v represented 0.06–0.1% of the total soluble protein in the transgenic tobacco leaves. Physiological experiments indicated that the control of phytochrome on developmental processes (e.g. light-induced germination) was strongly suppressed in the transgenic tobacco plants that produced the phytochrome-specific scF_v.

Tavladoraki *et al.* (1993) employed a similar genetic engineering method to construct a cDNA that encoded a scF_v derived from a monoclonal antibody that binds to the coat protein of the artichoke mottled crinkle virus (AMCV). The fidelity of their DNA construct and its expression of scF_v that specifically binds AMCV were first tested in bacteria. They then cloned the DNA construct into a plant expression vector and, following *Agrobacterium*-mediated transformation, obtained transgenic plants in which AMCV pathogenicity was very much reduced.

Firek *et al.* (1993) brought the system one step further. Their consideration was that retention in the cytoplasm of transformed cells will inhibit the accumulation of scF_v. They therefore used the signal peptide of the pathogen-related protein PRIa to secrete the phytochrome-specific scF_v outside of the tobacco cell. For that, they obtained the cDNA for the PRIa signal. This sequence was ligated upstream of the DNA that encodes the scF_v and the combined coding sequence was inserted into an appropriate plasmid vector. The latter was introduced into cells of an *Agrobacterium* strain and transgenic tobacco plants were obtained. Four out of ten analyzed plants showed production of scF_V. As anticipated, the scF_V was secreted into the apoplasm. From one of the transgenic plants, callus was induced. The callus was cultured to establish a cell suspension. Again, the scF_V was secreted into the medium and was found to bind to phytochrome. The addition of the PRIa signal peptide was indeed directing the scF_V to the outside of the cells and permitted the accumulation of relatively high levels in the leaf tissue (0.5% of total soluble protein). In the medium of the transgenic cell suspension the level of scF_V reached 5% of the protein in the medium (or about 0.25–0.50 μg/ml of culture medium). Further work from the same laboratory (Gandecha *et al.*, 1994) showed that the phytochrome-binding scF_V can also be conjugated to enzyme markers such as alkaline phosphatase.

Artsaenko *et al.* (1995) intended to test the effect of reducing the level of abscisic acid on the wilting of tobacco plants. They used monoclonal antibody specific for abscisic acid and accordingly engineered an scF_V-encoding DNA sequence. They added the code for the tetrapeptide Lys–Asp–Glu–Leu (KDEL) downstream of the scF_V coding region and found that when tobacco plants were transformed with an *Agrobacterium* vector that included the chimeric construct, the correct scF_V was formed in the transgenic plants and located in the ER. The transgenic plants that expressed the anti-abscisic acid scF_V had a greater tendency to wilt than control plants. This conforms with the known effect of abscisic acid deficiency of tobacco. But there was an enigma: the levels of abscisic acid in the scF_V-producing plants was several times higher than in the control plants.

Bruyns *et al.* (1996) also addressed the question of targeting the scF_V into either the cytoplasm or into the ER of the transgenic plant. These authors choose the signal peptide of an *Arabidopsis* seed storage protein and attached its DNA coding sequence upstream of the coding sequence of the scF_V that binds the human creatine kinase-MM. The overall production of scF_V with or without the signal peptide was rather low, at most 0.01% of the total soluble protein. On "hindsight", adding a signal peptide ahead of scF_v is much less efficient for the requested trafficking than to put the KDEL tetrapeptide downstream of the scF_v, to retain the scF_v in the ER. Probably, some navigations are maneuvered more efficiently by the "tail" than by the "head" (nature choose to anchor the protein "ships" to the ER by putting the KDEL at the aft rather than at the prow). A rather detailed study on the trafficking of scF_v in transgenic tobacco clearly indicated that targeting signals are very efficient tools for this trafficking (Schouten *et al.*, 1996).

A binational team consisting of Van Montagu's laboratory in Gent and R. Fischer from Aachen (De Jaeger *et al.*, 1997) explored the possibility to shortcut the need for monoclonal antibody production as a source of scF_v that will bind specifically to a defined antigen (dehydroflavonol 4-reductase of *Petunia*). They used the phage display library method to produce the requested scF_v. The main features of this method are the following. The DNA sequences that encode the variable domains of the requested antibody are fused to the code of the coat protein of a bacteriophage. Bacteriophages that contain this gene fusion are used to infect bacteria and, after plating, phages that express the variable domains and have the ability to bind the respective antigens are produced (a phage library). A few of these have the specific antigen-binding capability. From these, the antibodies or their subfragments are obtained. The phage display method to produce scF_v was exchanged by the PCR method in a subsequent study by Fischer and associates (Zimmermann *et al.*, 1998). In this later study, the investigators directed the scF_v to either the apoplast or the cytoplasm. For secretion to the apoplast, they attached the code for a mouse leader peptide upstream of the code for the scF_v. The accumulation of anti-TMV scF_v was several hundred times higher when directed to the apoplasm than when retained in the cytosol. Nevertheless, protection against TMV was much higher when the scF_v was retained in the cytosol. A further sophistication in using scF_v for protection against a pathogen was demonstrated by Fischer *et al.* (1999c). These authors intended to produce scF_v that has two specific binding capabilities. Such scF_vs are termed bispecific single-chain F_v fragments ($biscF_v$). $BiscF_v$s have potential advantages as therapeutic agents. They could, for example, crosslink cytotoxic T-cells to tumor cells. In the afore-mentioned study, it was first asked if plants and/or plant cells can indeed manufacture functional $biscF_v$ and if so, can they be useful to protect the respective transgenic plant against a viral pathogen. For addressing the last possibility, the authors intended to produce three kinds of $biscF_v$, all with the same two-binding capabilities but accumulating in one of three different locations: in the apoplast (i.e. outside the cells), in the ER or in the cytosol. They therefore constructed DNA sequences that encoded three $biscF_v$s (Fig. 10). In all three constructs, there was the same "core" construct. There was a code for scF_v24 that specifically binds to the neotopes of intact TMV virions and another code for scF_v29 that specifically binds to a cryptotope of the TMV coat protein monomer. The codes of two scF_v fragments of each $biscF_v$ were connected by the code for the cellobiohydrolase I linker of *Trichoderma reesii.* All three constructs also had, at their 5′ ends, the code for the untranslated leader of chalcone synthase (CHS). One $biscF_v$ construct had only the "core" DNA sequence. This was expected to express a $biscF_v$ ($biscF_v$ 2429-cyt) in the cytosol of the plant. A second construct had the same "core" DNA sequence but also a code for a murine leader peptide (LP), 3′ of the CHS code. This construct was aimed to express a $biscF_v$ ($biscF_v$ 2429-apoplast) that will be moved to the apoplast (i.e. outside of the cells). The third construct was as the second but also contained the code for KDEL at the 3′ end of the code for V_H29. This third construct was expected to express a $biscF_v$ ($biscF_v$ 2429-KDEL) that will be retained in the ER. The authors used the three constructs for *Agrobacterium*-mediated transformation using the *npt*II selectable gene (for resistance to kanamycin) in the respective vectors. They transformed tobacco leaf discs to obtain transgenic plants and also BY-2 tobacco cells to obtain transformed cell suspensions.

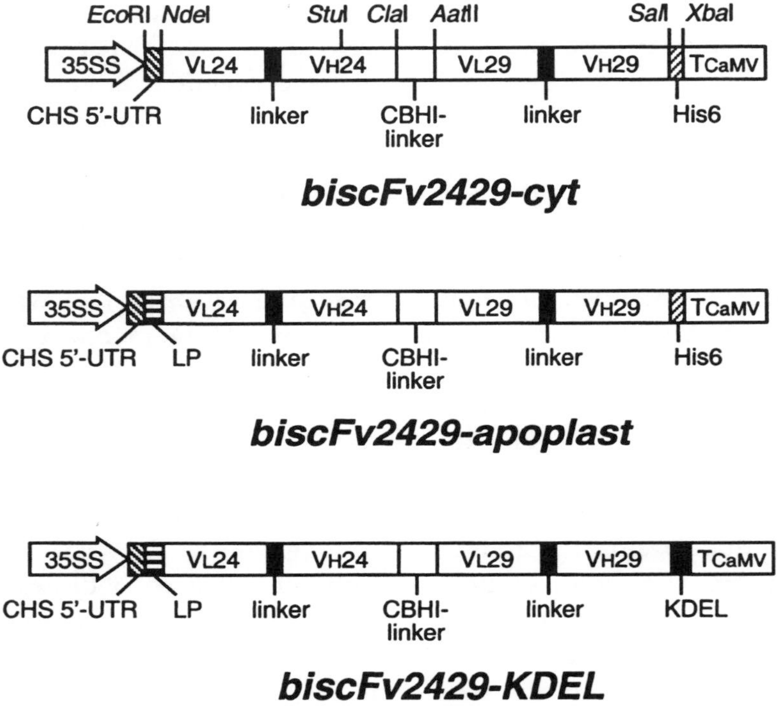

Fig. 10. Constructs for the expression of bisc F_v2429 in the cytoplasm, the apoplasm and the ER of plant cells. ScF_v cDNAs composed of mAb24 and mAb29, variable light chain (V_L) and heavy chain (V_H) domains were connected by an amino-acid linker. BiscF_V2429 was subcloned into the plant expression vector pSS to result in the cytoplasmic targeting vector biscF_V 2429-cyt; the apoplast targeting vector biscF_V2429-apoplast, and the ER retention vector biscF_V2429-KDEL. 35SS, enhanced CaMV-35S promoter; CHS-5′-UR, 5-UTR of chalcone synthase; LP, original mouse leader-peptide sequence form mAb24; His6, histidine-6 tag; KDEL, ER retention signal; TCaMV, CaMV termination sequence. (From Fischer *et al.*, 1999c.)

When the BY-2 cells were transformed with these constructs, the content of biscF_v 2429 was rather low. When transformed with biscF_v 2429-apoplast, the biscF_v was indeed found in the cell culture medium. When transformed with biscF_v2429-KDEL, the biscF_v2429 was detected only in the cell extract. It was also found that stably transformed

BY-2 cells produced biscF_v that had the expected dual-binding capability to intact TMV virions and to TMV coat protein monomers, and that the binding was specific.

When transgenic tobacco plants were regenerated after transformation with plasmids containing each of the three biscF_v constructs, the same kind of biscF_v distribution as in the transgenic cell suspensions was revealed. Thus, plasmids with biscF_v 2429-KDEL led to transgenic plants in which the biscF_v was retained in the ER. But the levels of production in the transgenic plants was several times higher than in the transgenic cell suspensions. The authors concluded that the highest levels of recombinant bispecific antibody were obtained in transgenic tobacco leaves that were transformed with the biscF_v 2429-KDEL-containing plasmid. This targeted the biscF_v to the ER lumen. Such a targeting has the advantage that the heterologous protein there could fold correctly, mature, and yield a high level of stable antibody. The authors concluded that the way for the production of specific biscF_v in plants is now open, and suggested that such biscF_v could serve for therapeutic purposes (see also Fischer *et al.*, 1999a; 1999b).

While directing scF_v to the ER will improve the correct processing of this fragment, the total amount of accumulation of this protein is limited. Moreover, the storage of leaves in which scF_v is accumulating is problematic. A scheme that shows cell compartmentation of Ig genes is shown in Fig. 11. Fidler and Conrad (1995) rightfully considered accumulation and storage of scF_v in seeds. For that, they engineered the respective DNA coding sequences. All the sequences encoded the scF_v that specifically recognizes the heptan oxazolone. To direct the expression to the seeds, the authors choose the *Vicia faba leB4* promoter and the *le*B4 signal. They also added a code for the *c-myc* tag at the 3′ end of the scF_v coding sequence to facilitate subsequent extraction and purification of the scF_v. When tobacco plants were transformed by *Agrobacterium* containing the above mentioned construct (with due additional genes in the plasmid), the authors obtained several groups of transgenic plants. Three groups of plants transformed with a construct that also encoded the *le*B4 signal peptide produced the requested scF$_v$ in their seeds. There were little differences in expression among these three groups which differed only in the base sequences of the 5′ end of the code for the scF$_v$ (i.e.: GCC ATC GCC GAT, ...GCC ACA ATG GCC GAT... or ...GCC GAT...). The expression was about 0.1 to 0.2% of total soluble protein with a maximum of 0.6% of total soluble protein in the seeds. The seeds could be kept for many months before extraction without impairment of the binding specificity of the scF$_v$. A further improvement in the methodologies to increase th level of stable accumulation of scFv in plant storage organs was reported by Conrad *et al.* (1998a, 1998b). Obviously, tobacco seeds are far from being an ideal "container" of scF$_v$; the seeds of some legumes such as soybeans and peanuts could be a better choice but the genetic transformation of these legume crops is still an "art" not shared by most investigators. The authors should therefore be credited for establishing the principle of producing scF$_v$ in seeds. The same authors (Conrad and Fiedler, 1998) recently reviewed the investigations, by themselves and by others, on the compartment-specific accumulation of immunoglobulins in transgenic plants.

Our next example could be included in this chapter or in the next chapter on antigens because it deals with the use of antibody employed as vaccines. This example deals with non-Hodgkin's lymphoma (NHL). NHL is a B-lymphocyte tumor that occurs infrequently in the human population. The disease can be suppressed by chemotherapy but complete cure is questionable. Moreover, chemotherapy is problematic and the prognosis is variable; remission after chemotherapy renders further therapy even more problematic. Each NHL patient has a unique cell-surface Ig. It is technically possible to use these unique IgGs for therapeutic purposes.

An immune response can be initiated in NHL patients when their specific tumor-surface Ig is rendered immunogenic by conjugation to keyhole limpet hemocyanin (KLH) and administered with an adjuvant. Such a treatment could improve the clinical outcome of

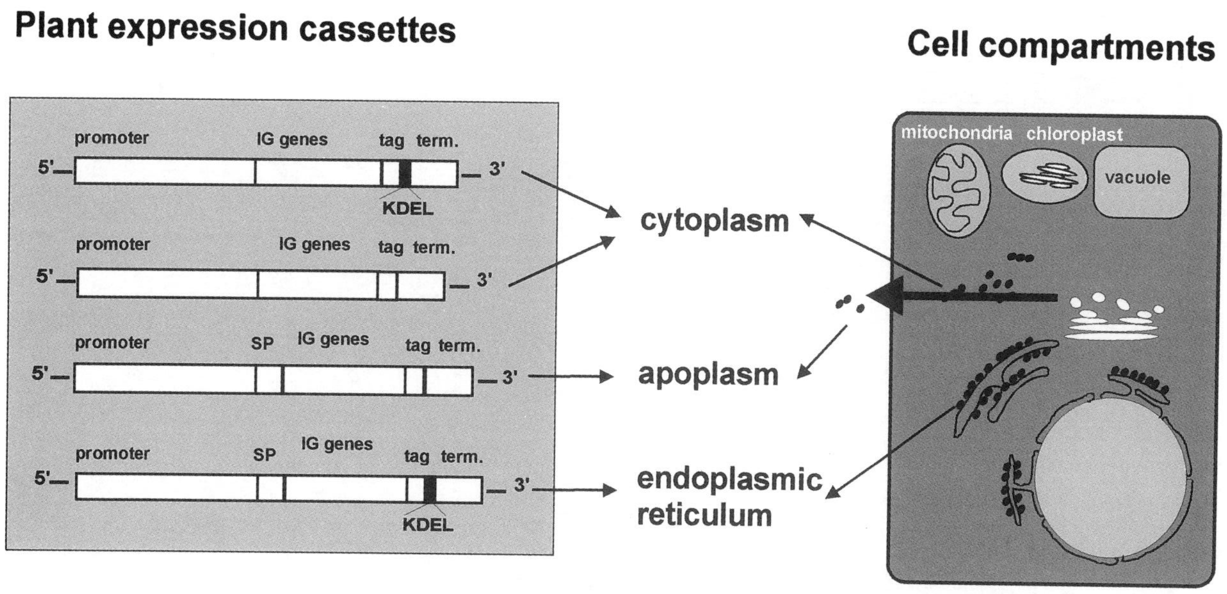

Fig. 11. Scheme of routing the immunoglobulins into cell compartments by different expression cassettes. (See text for abbreviations; courtesy of Dr. R. Fischer, Aachen, Germany.)

patients in chemotherapy-induced remission cases. But the production of such unique Ig in sufficient quantities and purity is rather difficult. The current procedure to produce Ig for patient therapy is based on the fusion of the patient tumor cells to a transformed human/mouse heteromyeloma cell. The hybridomas are then screened for secreted patient tumor-specific Ig and expanded for large-scale production of this protein. This procedure is problematic and time consuming, requiring many months for each case. For therapy, the single-chain variable region (scF_v) should be as effective as the full-size antibody.

The above mentioned considerations led McCormick *et al.* (1999) to look for a rapid production of tumor-derived scF_v. These nine investigators from the Division of Oncology, Stanford University School of Medicine and Biosource Technologies Inc., Vacaville, CA, chose transient expression in plants to produce the respective scF_vs. Transient expression was to be performed by inoculation of plants with a hybrid tobacco mosaic virus (TMV), in the genome of which the code for the specific scF_v has been inserted. In most plant hosts, TMV is a systemic disease. The virus replicates at the infection site and then the virions move from this site to other parts of the plants where additional viral replication takes place. Finally, most leaves will contain high titers of the virus. *Nicotiana benthamiana* is a host where systemic TMV infection takes place. The authors thus engineered a viral expression vector. This vector (Fig. 12) included the coding for a hybrid TMV–tomato mosaic virus. It also included the coding sequence for 38G13 scF_v. This scF_v represents the immunoglobulin from the 38G13 mouse B-cell lymphoma. For that, the coding sequence of an α-amylase signal peptide (SP) was put upstream of the scF_v code; this was expected to lead the expressed scF_v protein to the secretory pathway of the plant and promote its correct folding. The TMV/tomato mosaic virus amylase-scF_v-containing plasmids were transcribed *in vitro* to result in a 7 kb message. Approximately 2 μg of purified viral RNA was applied to a lower leaf of *N. benthamiana*. Viral infection symptoms occurred after 5–6 days and the leaves and stems of infected plants were harvested at about 12 days after

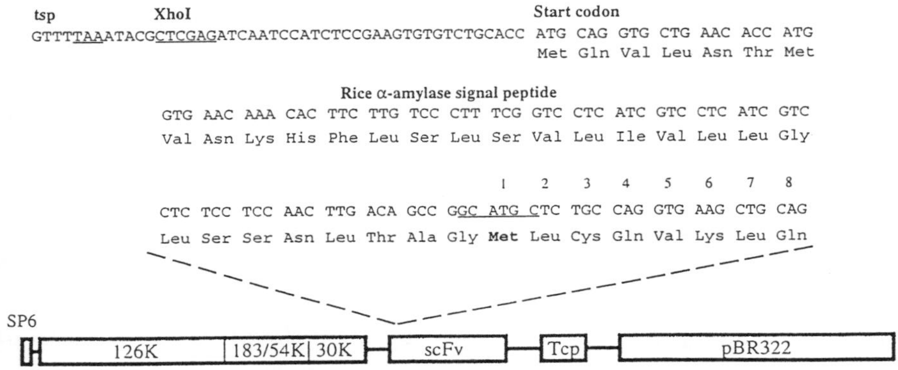

Fig. 12. Diagram of the viral expression vector used to produce murine 38C13 scF_v NHL antigen in *Nicotiana benthamiana* host plants. Shown are the sequences encoding the rice α-amylase signal peptide, fused inframe to the *Sph*I restriction site (underlined) introduced by PCR, 5′ of the 38C13 heavy chain. Also indicated in the linear map are the positions of the SP6 transcription start site, the 126-kDa, 183-kDa and 30-kDa proteins from TMV, and the tomato mosaic virus coat protein (Tcp) as well as pBR322 sequences for bacterial propagation. (From McCormick *et al.*, 1999.)

infection. The content of the intercellular spaces was collected by first infiltrating the plant material with a buffer followed by mild centrifugation and filtering through a nylon screen. This filtrate was passed through 0.2 μm pores of a membrane and stored (at −80°C). This latter filtrate (IF) was thawed on ice to provide the material for further purification as well as for analyses and vaccination of mice. It was found that the 33-aa signal peptide was translated and the translation proceeded into the 38C13 scF_v. In the plant, this signal was cleaved off between the last Gly of the signal peptide and the first Met of the scF_v. Several tests indicated that the required 38G13 scF_vs were produced by the virus-infected plants. The scF_v seemed to be stable in plants, at least up to 14 days after inoculation. The calculated levels of scF_v reached 62 μg/ml or 30 mg per kg of the wet weight of leaves. Most of the scF_vs were properly folded and specifically recognized the respective anti-idiotype. Also, the plant-produced scF_v

had no interfering glycosylations. The investigators then questioned if the plant-produced 38G13 scF_v can elicit an anti-38G13 response and consequently protect mice from 38G13 tumor challenge. They thus vaccinated mice with 38C13 scF_v, with or without an adjuvant. Such a response was detected in the mice after the second and third vaccinations. The addition of adjuvant increased the response and initiated an IgG2a isotype response (antibody of the class IgG2a have been correlated with augmented anti-tumor protection). When mice were first vaccinated with plant 38G13 scF_v (with the adjuvant), they were protected (up to 90% survival after 60 days) from subsequent injection of 38G13 tumor cells. All nonvaccinated and tumor-cell injected mice died within about 15 days. The authors provided convincing evidence that the plant-produced 38G13 scF_vs exist as monomers and dimers in the IF and that these scF_vs, rather than any other components of the IF, are responsible for their anti-tumor effect. They suggest that this procedure can be applied to treat human NHL patients.

A general update on scF_v antibodies expressed in transgenic plants with detailed protocols was provided by Conrad *et al.* (1998a).

3.8. Production of Full-Size Antibody by Transgenic Plants

3.8.1. *Studies on the functionality of antibodies produced in transgenic plants*

The first successful production of full-size antibodies was published soon after *Agrobacterium*-mediated genetic transformation became a routine procedure. Hiatt *et al.* (1989) adopted the strategy of obtaining two kinds of transgenic plants, each producing either the heavy chain or the light chain of the immunoglobulin. Sexual crossing of plants containing the heavy chain with plants containing the light chain should then result, among the respective progeny, in plants producing both chains. It was anticipated that the latter transgenic plants will produce full-size and functional antibodies. To test the validity of this strategy, the authors choose a mouse monocolonal antibody (IgG_1) that binds specifically to a phosphoronate ester (6D4). cDNAs were obtained for the γ (gamma, heavy) chain and for the κ (kappa, light) chain. Each was inserted, either with or without a leader, to a transformation vector, and the vector was used in *Agrobacterium*-mediated transformation of tobacco. Thus, four groups of transgenic tobacco plants were obtained: encoding the gamma chain with leader, encoding the gamma chain without leader, encoding the kappa chain with leader and encoding the kappa chain without leader. The constructs without leader led to the production of very low levels of the respective chains. Then, plants producing the gamma chain were crossed with those producing the kappa chain. Some plants of the progeny of this cross yielded full-size antibody that showed the expected antigen-binding capability. Although the authors assumed that the assembly of the two chains takes place in the ER, no signal to retain the chains in the ER was added to the transformation construct.

Düring *et al.* (1990) used a different strategy for the same goal: production of full-size and active antibody in transgenic tobacco plants. These authors included the cDNAs for the two chains of a monoclonal antibody (B1-8) in one construct. Each cDNA was preceded by a different promoter. They also added the code for a signal peptide of α-amylase, upstream of the coding sequence of the light chain. The results indicated that the two chains were synthesized and assembled in transgenic tobacco plants. This was shown by several methods (e.g. affinity chromatography, immunoblotting, "tissue-printing" and electron microscopic immunogold labeling). The immunogold labeling indicated that the assembled antibodies were located in the ER but ("surprisingly") also inside the chloroplasts at the thylakoid membranes. Again, as in the previously mentioned study, the DNA construct did not contain a coding sequence of a signal (e.g. KDEL) for retention in the ER.

De Wilde *et al.* (1996) looked at the exact localization of a full-size Ig and its F_{ab} fragments that were produced in transgenic plants and were destined to pass through the ER towards the cell wall. This IgG has a specific binding to the human creatine kinase. The light and heavy chains had a seed-storage-protein signal peptide (2S2) that first led them to the ER where they were probably processed into functional IgG and then were moved to the cell wall. Immunogold labeling clearly indicated that although the full-size IgGs had a molecular mass of 146 kDa, they were secreted out of the cell wall and into the intercellular spaces of the transgenic plant (*Arabidopsis thaliana*) in which they were produced.

The most abundant immunoglobulin in mucosal secretions is the secretory immunoglobulin A (SIgA). It is commonly composed of two IgA monomeric antibody units that are associated with a small joining polypeptide (J) and with a secretory component (SC). The final synthesis of the chimeric SIgA *in vivo* is an elaborated process. Ma *et al.* (1995) thus intended to assemble the SIgA in transgenic plants. For that, they engineered an elaborated set of cDNAs. Each of these constructs contained one coding sequence. One cassette contained the codes for the H chain and the L chain of the murine antibody (Guy's 13). Another cassette contained a code for a J chain and the third cassette contained the code for the SC. After the cassettes were cloned into appropriate transformation vectors, they were used in *Agrobacterium*-mediated transformation of tobacco plants. The authors also exchanged the coding sequence for the Cγ domains with that for Cα domains in order to result in an IgA–G (rather than the original IgG1). Three groups of transgenic plants were obtained: plants producing IgA–G, plants producing the J chain and plants producing SC. When the plants with IgA–G were crossed with those producing J chains, they obtained in the progeny of this cross an Ig that was approximately twice the size of IgA–G (~ 400 kDa). Transgenic plants with this apparent double IgA–G were then crossed with a homozygous transgenic plant that produced SC. Consequently, the authors obtained plants that expressed the anticipated SIgA–G (470 kDa). These antibodies had the required specific antigen-binding capability and the yield was 200–500 μg per gram of fresh-weight leaves.

Van Engelen *et al.* (1994) reported on the research of a dozen investigators from three institutes in the Netherlands that aimed to express functional antibody in the roots of transgenic tobacco plants. The strategy of these investigators was to include the cDNAs of both the heavy chain and the light chain in the same transformation vector. Each of the cDNAs had its own promoter. The heavy-chain cDNA was driven by a 35S CaMV promoter (with enhancer) while the light-chain cDNA was under the control of the TR2′ promoter (this promoter was claimed to be root specific). The cDNAs were obtained from a hydridoma that produces antibodies against a cutinase that is produced by the fungal pathogen *Botrytis cinerea.* To obtain the final cDNAs, the investigators used PCR procedures and introduced some modifications at the 5′ end of the coding sequence for the immunoglobulin chains. The two cDNAs, with their respective promoters and terminators, were introduced into the *Agrobacterium* vector plasmid in either of two orientations: in tandem (MA) or in divergence (MB) with either about 1,700 or 1,300 base-pair distance, respectively, between the genes. Transgenic tobacco plants were obtained that produced the immunoglobulin chains. Most of the Igs were apparently representing full-size (~ 150 kDa) antibody but there were also Igs that represented $F(ab')_2$-like antibody. There was not only antibody production in the roots but also in the other plant organs. The antibody did recognize the cutinase and more antibodies were produced in transgenic plants transformed with the MB gene orientation than with the MA (tandem) orientation. The main achievement of this research was to show that the codes for the two immunoglobulin chains can be included in one transformation cassette. The maximal accumulation of functional antibody was about 1% of total soluble protein. This research did not reach the evaluation of resistance, in the transgenic plants, against the fungal pathogen. No wonder, *Botrytis cinerea* is not a pathogen of tobacco roots. Actually, recent molecular taxonomy indicated that *Botrytis,* as other oomycetes, is not a fungus at all but an alga.

The question of protection against a pathogen in transgenic plants that produce a specific antibody was handled by Voss *et al.* (1995). In this study, the monoclonal antibodies that recognize specifically the surface of intact tobacco mosaic virus (TMV) virions (mA624) were utilized. The cDNAs for the heavy and light chains were cloned and the murine 5′ untranslated leader was replaced by a plant 5′-UTR of chalcone synthase. The code for the translated murine leader was placed upstream of the chain code. The cDNAs of the chains were flanked with the necessary plant promoters (e.g. the 35S CaMV promoter in which a part was duplicated) and terminators. These were introduced into an *Agrobacterium* vector plasmid in the tandem orientation. After the *Agrobacterium*-mediated transformation, transgenic plants that produced the transcripts for the immunoglobulin chains were obtained. Among these, there were plants that produced full-size antibody which recognized the TMV surface. When the progeny (selfed) of plants that produced such an antibody were tested for protection against TMV infection, it was found that there was a reduction of infection (of between 24 and 68%) but not immunity to TMV. The investigators concluded that the bottleneck of TMV immunity is probably a relative lack of heavy chain production. It should be noted that the correct folding and assembly of functional antibodies take place in the ER. In this study, the IgG had a leader that secreted the antibody to the intercellular spaces (where they should meet the TMV virions) but no signal to retain the antibody fragments in the ER were included in the transformation cassette. Another point should be recalled. The phenomenon of the silencing of transgenes, as discussed in Chap. 2, was not yet considered crucial in 1995. Especially, silencing related to several 35S promoter sequences in the same transgenic plant cells could reduce the accumulation of the alien protein (i.e. IgG) in these cells.

Another approach to reduce parasitism in transgenic plants is represented in the report of Baum *et al.* (1996). These investigators asked whether or not antibodies ($6D_4$) that recognize specifically the stylet secretions of the root-knot nematodes (*Meloidogyne incognita*) will reduce nematode parasitism in transgenic plants producing such antibody (they termed antibody produced in transgenic plants: plantibody). To test this possibility, these investigators included either the cDNA for the heavy chain (of the monoclonal antibody $6D_4$) or the respective cDNA for the light chain in two separate transformation vectors. The parasitism of root-knot nematodes consists of an elaborate action/response system between the nematodes and the root cells. One component of this system is a secretion from the nematode esophagal glands, that induces a change in plant cell division and metabolism. The affected plant cell undergoes a series of nuclear divisions without cell division to form a giant feeder cell that nourishes the nematode. A monoclonal antibody ($6D_4$) was previously produced that specifically recognizes the stylet secretion of the root-knot nematode. The two transformation vectors were used to transform separately tobacco plants via *Agrobacterium* infection. The authors observed that in the plants transformed with the vector encoding the heavy chain, the latter usually formed a dimer (of about 97 kDa) that was relatively stable in the transgenic plant. The light chains in the other group of transgenic tobacco also included the leader peptide (16 amino acids). The progeny of crosses between heavy-chain- and light-chain-producing transgenic tobacco were analyzed and plants that produced both chains were further tested. It appears that the level of functional antibodies in the plants was low: about 0.01% and 0.0003% of total soluble protein in leaves and roots, respectively. As the promoter for the cDNA of the Ig chains was the 35S CaMV, it should have been no surprise that low levels of the chains were revealed in the roots (the location of nematode/plant-cell interactions). It should also be noted that the plantibody in the plant existed as IgG-like monomers rather than IgM pentamers. While these monomers still maintained their specific binding to the nematode-derived antigen, it was found that transgenic plants expressing these antibodies were not protected against nematodes. The authors suggested that the plantibodies were mainly secreted into the intercellular spaces. For effective elimination of the nematode-derived secretions, the antibodies should have been retained in the cytoplasm. The authors thus recommend to produce in the future scF_v that will be retained in the cytoplasm, into which the nematodes inject their secretion.

Wongsamuth and Doran (1997), who worked in Sydney, Australia, looked at the cultured roots as a source of functional antibody. They intended to exploit the characteristics of roots derived from *Agrobacterium rhizogenes* infected tobacco to maintain perpetual growth in culture (see Secs. 2.5 and 3.8.2). The authors actually utilized a transgenic tobacco line, produced previously by the team of J.K.-C. Ma, A. Hiatt and associates (Ma *et al.*, 1994). This line expressed Guy's 13 antibody. This antibody recognizes a surface protein of the dental pathogen *Streptococcus mutans* and thus has potential use to prevent dental caries. Seedlings of this transgenic tobacco lines (provided by A. Hiatt) were transformed with a strain of *A. rhizogenes* and the resulting hairy roots were grown in liquid culture. As control, the roots were used to initiate callus and from that callus, cell suspensions were derived. The ability of the cultured hairy roots to produce antibody was evaluated. Eight out of 17 such root cultures were found to produce antibodies. Antibodies were found in the root tissue as well as in the culture medium. There was increase in antibody production up to a number of days (about 15 days in the tissue and 18 days in the culture medium). The type of medium had an effect on the longevity of the root cultures (the B_5 medium was better than the MS medium) and some additions to the culture medium (polyvinylpyrolidone, gelatin) improved the secretion of antibody into the medium. The authors related at least part of the decline in functional antibodies to their proteolytic degradation. They found that both shake culture (started with 10 g fresh weight of roots in 50 ml medium in 250 ml flasks) as well as in 2-liter bioreactors (percolated with humidified air) were suitable to obtain antibodies in tissue and culture media. The presence of antibody was evaluated by ELISA methods and by Western blot hybridization to mouse IgG_1. The functionality of the antibody was tested with root homogenates that specifically agglutinated *S. mutans* but not *E. coli* . The maximal level of antibody in the medium was 18 mg per liter (or 1.8% of total soluble protein). As in some of the previous studies in this area, the construct was not built to express the antibody specifically in the roots. Possibly in the future, such specific constructs that not only cause expression in roots but where the production can be initiated by a specific inducer of the respective promoter will improve the biotechnology of antibody production in hairy-root culture. With a smile, we may add that if this will lead to hairy-root cultures of carrot, there is a potential for both tooth and eye care by such transgenic roots.

3.8.2. *Plantibodies for therapeutic purposes*

The notion that *plantibodies* (antibodies produced in transgenic plants) can serve for therapeutic purposes was brought forward since the first *plantibodies* became available. The first study that not only suggested such a possibility but went ahead to work on a plausible system that should lead to passive immunization against a malady was carried out by Ma *et al.* (1994). This study intended to furnish antibodies for topical immunotherapy. The research was based on the availability of a monoclonal antibody (IgG1) that recognizes a 185 kDa cell surface of *Streptococcus mutans* (Guy's 13, already mentioned in Sec. 3.8.1) and on information that a similar monoclonal antibody can prevent colonization of *S. mutans* and thus protects against dental caries in *nonhuman primates* (an elegant term for apes). Moreover, application of Guy's 13 to the teeth of human volunteers prevented *S. mutans* implantation. This binational team of investigators (J.K.-C. Ma and T. Lehner of Guy's Hospital in London, and A. Hiatt and associates of the Scripps Research Institute in La Jolla) used Guy's 13 to construct the respective cDNAs for the heavy and the light chains. They also constructed two hybrid coding sequences for the heavy chain so that parts of the constant domain of the γ heavy chain code were exchanged with codes of the constant domains of the heavy chain of a murine IgA (MOPG315). Three types of heavy chains should thus be translated:

G13: Var – $c\gamma_1$ – $c\gamma_2$ – $c\gamma_3$
G1/A: Var – $c\gamma_1$ – $c\alpha_2$ – $c\alpha_3$
G2/A: Var – $c\gamma_1$ – $c\gamma_2$ – $c\alpha_2$ – $c\alpha_3$

The coding sequence for one of the three heavy chains was introduced into each of three plant transformation vectors that also contained a coding sequence for a mouse immunoglobulin leader. The coding sequence for the light chain was also introduced into each of the three plant transformation vectors. *Agrobacterium*-mediated transformation with these four vectors resulted in plants that were analyzed by ELISA for the production of the respective heavy or light chains. Plants producing the light chain were crossed with plants producing G13, G1/A or G2/A heavy chain. The progeny resulted in three plants with light chain and G13 heavy chain, four plants with light chain and G1/A heavy chain and three plants with light chain and G2/A heavy chain. The existence of the chains in the plant extracts was verified by Western blot hybridization under reducing conditions. Western blot analyses under nonreducing conditions was also performed to verify the assembly of the chains into the respective antibodies. Bands showing correct assembly (150 kDa) did appear but, in addition, bands representing smaller immunoglobulins were detected. The plantibodies did indeed recognize the SA I/II antigen and were bound to the *S. mutans* cell surface. Incubation of the plantibodies with a bacterial culture showed aggregation of *S. mutans*. There were no gross differences in the binding capability between the three types of plantibodies but it was also established that plantibodies with a chimeric heavy chain are functional.

In a further study by this team (Ma *et al.*, 1998), which also included investigators from a biotech company, a functional comparison was made between a monoclonal secretory antibody produced by transgenic plants and the murine IgG from which the plantibody was derived. The plantibodies used in this study were the SIgA/G secretory antibodies produced previously in transgenic tobacco (Ma *et al.*, 1995). First, the binding affinity of SIgA/G to the *S. mutans* antigen I/II (SA I/II) was similar to that of the murine IgG Guy's 13. The functional affinity was fourfold higher for the SIgA/G than the murine antibody. The investigators then used human volunteers to test the survival of the antibodies in the oral cavity. The murine antibodies disappeared after one day while the SIgA/G survived for three days. Further, human volunteers who had *S. mutans* in their teeth were subject for a more decisive test. First, the contamination of *S. mutans* was reduced to a stage at which no bacteria could be detected. This was performed with repeated topical application of chlorhexidene gluconate for nine days. Then, the volunteers were divided into groups receiving either control treatments or antibodies. For those receiving control treatments, all had recolonization of bacteria at the latest after 58 days. Those who received either Guy's 13 IgG or SIgA/G (two applications per week, for three weeks) had no recolonization after 118 days. The plantibodies and the Guy's 13 IgG had no effect on the recolonization of another bacterium (*Actinomyces* sp.). No side effects were revealed and it was suggested that the purified plantibodies are not immunogenic when applied orally (see also: Ma and Hein, 1995; 1996; Ma *et al.*, 1997; Ma and Vine, 1999).

We shall make a shift from dental infection by bacteria to vaginal infection by a virus. The latter infection was handled in a more recent publication by a mixed team of commercial research laboratories (ReProtect, Baltimore, MD; Agracetus, Middleton, WI; Protein Design Labs, Mountain View, CA) and Johns Hopkins University, Baltimore. These researchers (Zeitlin *et al.*, 1998) investigated plant-produced anti-herpes-simplex-virus-2 (HSV-2) antibodies. The description of their experimental procedures is rather brief but they based the construct that encoded the antibody for plant transformation on a monoclonal "humanized" IgG that binds the glycoprotein B of HSV. The codes for the heavy and the light chains were inserted into an identical cassette in which the 35S CaMV promoter was used. A code for the plant signal peptide, tobacco *extensine*, was added upstream of the chain codes and the *Nos* terminator was added downstream of these codes. The investigators also added a codon for methionine to the 5′ end of the mature heavy chain and also made a change to the 5′ of the light chain. Soybean plants were transformed by the biolistic method in which plasmids with the heavy-chain transformation cassette and plasmids with the light-chain cassette were mixed before bombardment. The plant IgG was purified and compared to mammalian IgG. The investigators found that the plantibodies were capable of neutralizing HSV-2 *in vitro* just as the mammalian antibody. Both types of antibody reduced the HSV-2-induced cytopathic effect (CPE) and a 100% reduction was achieved by 2 μg/ml of plantibodies. Moreover, the plantibodies showed stability in human cervical mucus as well as in human semen, at least as the stability of the mammalian antibodies. When a mouse model was used for vaginal HSV-2 transmission, the investigators found that both types of antibodies had a similar effect of protection against vaginal inoculation of HSV-2. Full protection was achieved by a plantibody concentration of 1,000 μg/ml (i.e. 10 μg per application). The ability of the plantibody to neutralize HSV-2 in the acidic conditions of the human vagina and to maintain their activity for a day or longer is an important result of this study. It opens additional possibilities as to the production of plantibodies to eliminate other infections in the human vagina.

3.8.3. *Transient expression of antibody in transgenic plants*

There are cases in which large quantities of a specific antibody are required and should be available as quickly as possible. To produce transgenic plants which express a transgene that was stably integrated into their respective genomes is a long process. The alternative is transient expression. The process of transient expression in plant leaves has been available for many years and it can be facilitated by agrobacteria. A group of German investigators (Vaquero *et al.*, 1999) intended to test the large-scale transient expression of full-size antibody and scF_v for therapeutic and diagnostic purposes. They choose the antibody against the human carcinoembryonic antigen (CEA). CEA is a cell surface glycoprotein that is found in human colon and pancreatic cancers, as well as in about half of breast cancer cases and in other tumors of epithelial origin. Thus, antibodies against CEA have a two-fold application. Anti-CEA antibody may be used for antibody-mediated cancer therapy and for *in vivo* imaging of tumors. The authors focused on a specific antibody, T84.66, that binds to the A3 domain of CEA with high specificity and affinity. Such chimeric T84.66 recombinant mouse/human antibody was available, and the respective antibody fragment, scF_v T84.66, was also produced previously. The question was whether such an antibody and scF_v can be manufactured in large quantities in plants by transient expression.

These investigators thus engineered three constructs; one for the transient expression of the scF_v T84.66 and two constructs for the expression of full-size T84.66 antibody. For the first construct, they engineered the chalcone synthase 5′ untranslated region and a plant codon-optimized leader, upstream of the coding sequence for the scF_v T84.66. This coding sequence was followed by a His6 tag to facilitiate future extraction of the scF_v. The code for the 3′ untranslated region of TMV RNA was inserted downstream of the His6 tag. The two constructs for the full-size antibody contained at the 5′ the omega leader enhancer (from TMV) and one of them also contained the coding sequence for the chimeric light chain of T84.66 (variable light chain of mouse; constant light chain of human). The other construct contained the coding sequence for the chimeric heavy chain (variable heavy chain of mouse and C_H1, C_H2, C_H3 of human). The code for the KDEL motif for ER signaling was added downstream of the coding for the heavy chain. The two constructs for the full-size antibody were terminated with the untranslated region of TMV RNA. The constructs were engineered into *A. tumefaciens* plasmids.

For transient expression, detached tobacco leaves were infiltrated with a suspension of *A. tumefaciens* that harbored the respective plasmids. For transient expression of the full-size T86.44 antibody, the two agrobacteria, one with the code for the light chain and one with the code for the heavy chain, were mixed before infiltration. After infiltration, the leaves were maintained for about 60 hours in a closed container and then either frozen (−80°C) and later extracted or extracted right after incubation.

After due extractions and purifications, it became evident that the transient expression resulted in biologically active scF_v and full-size antibody. The authors calculated that the yield of the full-size T84.66 antibody was approximately 1 mg per kg of leaves. They also estimated that for therapeutic treatment of colon tumor in the USA, between 6.5 and 130 kg of antibody is required. Taking the lower-range quantity of antibody (6.5 kg), this will require the infiltration of 6.5×10^6 kg (6,500 tons) of leaves. Quite a lot of tobacco leaves but then there are about 650,000 new colon cancer patients diagnosed each year in the USA. Let us extrapolate, with a smile. If we take the upper limit (130 kg T84.66 antibody) and/or the need of tobacco leaves to transiently express other therapeutic antibodies, the tobacco industry could shift completely from cigarettes to antibody production. This could have a twofold impact on human health. For additional information on the studies of Fischer and associates, see Fischer *et al.* (1998; 1999a; 1999b; 1999d).

Chapter 4

Antigens

4.1. Introduction

At first sight, the induction of the adaptive immune system by plant-derived antigens is looked upon as a simple and cheap solution for vaccination against microbial and viral parasites. As it was considered to be much cheaper to produce antigens in transgenic plants than by other means (e.g. human cell cultures), the concept was developed such that plant-derived antigens will be a solution to combat several severe maladies in developing countries where antigens produced by traditional means will be prohibitively expensive. This concept was noted repeatedly in the many reviews on the production of antigens by transgenic plants that were published since 1995 (e.g. Moffat, 1995; Richter *et al.*, 1996; Arntzen, 1998; Liu, 1998; Mor *et al.*, 1998; Ma and Vine, 1999; Richter and Kipp, 1999).

As we shall see below, the "first sight" may be misleading. We still do not know what is the exact (and minimal) antigen required to trigger the adaptive immune system against most human parasites. For oral vaccination, an antigen that is produced inside a foodstuff, such as fruits and tubers, is acceptable even without further purification. But for injections, the antigen should be purified to a clinical acceptable level. The problem is that in most foodstuff candidates such as bananas, tomatoes and potato tubers, the level of protein is low and the antigen commonly constitutes only a fraction of one percent of the protein.

It may come as a surprise to some biologists but the only real information on how a three-dimensional structure evolves from a linear DNA sequence lies in the viral coat proteins. Such real information does not exist in any cellular organism. Schematically, the information in sequences of RNA or DNA for the three-dimensional structure of viral envelope is commonly in encoding a polyprotein. The latter includes one or more subunit proteins and one or more specific proteases that have a role in cutting the polyprotein into individual coat proteins. The latter will then self-assemble into virus-like particles in which the viral genome will be included. When biologists speak about genes encoding a certain *shape* of a leaf or a flower, etc., they do not mean that this gene has all the information of how to construct the leaf or the flower, but rather that this gene *affects* the shape, changing it from one form to another. As for antigenicity, in some cases, it is known which components of the coat protein of a virus will trigger the adaptive immune system (as in hepatitis B virus, HBV); while in other viruses, it is not clear which components of the virus coat are essential for this triggering (as in hepatitis A virus, HAV). Moreover, unless protected, such as by encapsulation, to avoid degradation (high pH in the stomach, proteolytic degradation) until the antigen reaches its carriers, the M cells of the gut, relatively high doses of antigen are required for successful oral vaccination. Furthermore, application of the same antigen at different, specific locations (i.e. of the gut) will elicit different mucosal immunogenic responses. One estimate is a 100-fold higher oral dose than the dose required for parental injection. If we speak about antigen-containing banana, we can make a very rough estimate of the quantity of bananas required. Let us assume that 100 μg of antigen is required for one oral vaccination and that the antigen reaches 0.01% of the protein in the transgenic banana fruit. This means that 1 g of banana protein is required. As bananas contain about 1% protein, one oral vaccination will require 100 g of banana; hence, one medium-size banana. But, this is a very rough estimate and was merely provided here to give a notion on the food quantity that is required.

Then, comes the commercial consideration. Producing inexpensive vaccines for poor countries is a nice approach but "someone" has to pay for the development of such vaccines. The time required from the conceptualization to the clinical trials of plant-derived vaccines is probably about ten years. The last stages of this development are very expensive, much beyond what can be invested by public research institutions (e.g. universities). Hence, commercial companies have to be recruited. The latter will invest in development only when they see a very good chance to regain their investment. Moreover, small commercial companies may not survive the many years of investment. For the very large international drug companies, we should take into account the attitude of their CEOs. Some of the latter may prefer to donate a modern hospital to a developing country, which within two years will be operational (and carry the name of the donating company) rather than to invest an undefined amount of money for many years in a project with a high risk and expected low profit.

As we shall see, when describing specific research attempts to produce effective antigens in transgenic plants, there are other problems that have to be solved. Thus, while some success is emerging, there are not many effective antigens produced by plants that are available today.

Clearly, plant-derived vaccines are based, in each case, on previous research in medical and veterinary virology, immunology and vaccinology. A very vivid historical account on these fields of endeavor (with interesting personal experiences) was recently compilated by M.R. Hillman (1999). This article surveys 60 years of research activity and is highly recommended to those interested in the history of vaccinology.

We shall divide this Chapter into two parts. For the first part, we shall handle the production of antigens in transgenic plants that were stably transformed (i.e. by *Agrobacterium*-mediated transformation). For the second part, we shall handle the production of subantigen polypeptides that are integrated into the coat protein of plant viruses and are exposed by them as antigenic epitopes. The plant virus is used in plant inoculation and spreads systemically (but transiently) in the plant.

4.2. Antigens Resulting from *Agrobacterium*-Mediated Transformation

In the introduction to Chap. 3, we narrated the original vaccination by Edward Jenner in 1796 (published in 1798). But according to Margaret A. Liu (1998), active immunization against smallpox was performed much earlier by "the ancient Chinese". This was done by *variolation,* in which a small quantity of scab from a lesion of a smallpox-infected person is intranasally inoculated. Lady Mary Montagu observed variolation in Turkey in the early 1700s. Was the variolation procedure transported along the Silk Road (in parallel or independent of smallpox), or, perhaps the procedure and/or smallpox arrived with the Mongol warriors? M.A. Liu did not tell us. Whatever, intranasal vaccination seems to justify more attention in the future. It is obviously an interesting port of entry into the mammalian body.

4.2.1. *Antigens from pathogenic viruses*

4.2.1.1. *Hepatitis B virus*

The first report on the production, in transgenic plants, of an antigen derived from a pathogenic virus came from Texas, USA (Mason *et al.*, 1992). The team included H.S. Mason, D.M.-K. Lam and C.J. Arntzen, then at AgriStar Inc., LifeTech Industries and Texas A & M University, respectively. These investigators moved, in subsequent years, to other locations and also changed their research direction to produce antigens for pathogenic bacteria. In their 1992 publication, they intended to demonstrate the feasibility of the production, in transgenic plants, of a vaccine that will have immunogenic properties. They chose an important human pathogen: hepatitis B virus (HBV). The ultimate goal of this team was to produce an oral vaccine against HBV in transgenic plants. They based their study on the already commercially available vaccine against HBV produced in yeast cultures. The idea was to use transgenic plants in order to reduce the production costs of this vaccine. The recombinant, yeast-derived vaccine contains the hepatitis B surface antigen (HBsAg). This antigen can self-assemble into subviral particles that bind HBV antibodies. The question these investigators asked was whether or not HBsAg produced in transgenic plants will also self-assemble and consequently bind to monoclonal antibodies directed against human-serum-derived HBsAg.

The cDNA encoding the HBsAg was obtained from a plasmid (pMT-SA). The investigators thus constructed plasmids (pHB101 and pHB102) that contained the left border (LB) and the right border (RB) of an *Agrobacterium* T-DNA (binary plasmid). The plasmid also included the kanamycin resistance, selectable gene (*npt*II) that was flanked by the *Nos* promoter and the *Nos* terminator. The cDNA of HBsAg was flanked by the *Nos* terminator and either the regular CaMV 35S promoter (pHB101) or the CaMV 35S promoter with a dual enhancer (pHB102). The latter plasmid also contained an untranslated tobacco etch virus (TEV) leader. Either of these latter plasmids was used in *Agrobacterium*-mediated transformation (with the strain LBA 4404 that contains a helper plasmid). Tobacco leaf sections (*N. tabacum* cv Samsun) were transformed by the conventional method and the regenerating shoots were selected on kanamycin medium. Shoots were then rooted and the resultant plants were used for different analyses. Northern blot hybridization revealed the HBsAg transcript. When the vector contained the 35S CaMV promoter with the dual enhancer (pHB102) and TEV leader, the transcript was slightly larger than in plants transformed with the pHB101 plasmid (regular 35S CaMV promoter). The HBsAg protein, estimated by binding to HBsAg monoclonal antibody, was several times more abundant in plants transformed with the pHB102 plasmid than in plants transformed with the pHB101 plasmid. Moreover, the HBsAg was capable of self assembly into particles having an average diameter of 22 nm. This size is comparable to yeast-derived HBsAg particles.

This study thus showed that HBsAg which binds to anti-HBsAg can be produced in stably transformed tobacco plants. The level of antigen was rather low, about 0.01% of the soluble protein in the leaves.

This study was continued and resulted in a further report (Thanavala *et al.*, 1995). In the latter work, the authors investigated the immunogenic response using the recombinant hepatitis B surface antigen (rHBsAg) that was purified from the leaves of transgenic tobacco. It was found that the anti-hepatitis B response (in mice) to the tobacco-derived rHBsAg was similar to the response after immunization with the yeast-derived rHBsAg. When mice were injected three times with either the tobacco- or the yeast-derived rHBsAg, the antibody peaked at about seven weeks after the last injection. Furthermore, the tobacco-derived rHBsAg elicited IgM antibodies as well as all the IgG subclasses, while the yeast-derived rHBsAg induced primarily IgG1 and IgG2 (but very little IgG3 and IgM). Mouse T-cells could be stimulated by the tobacco-derived rHBsAg. BALB/c mice were first primed with the tobacco rHBsAg. The T-cells of these mice were then used in an *in vitro* test to reveal if these T-cells could be stimulated by the tobacco rHBsAg, by the yeast-derived rHBsAg or by other stimulators. It was found that, indeed, the tobacco-derived rHBsAg stimulated the T-cells (evaluated by [^{3}H] thymidine incorporation) to similar levels as the stimulation by the yeast-derived rHBsAg. No stimulation of these (*in vitro* cultured) T-cells was observed after control triggering. The authors suggested that, in future work, HBV antigens may be produced at a higher level, possibly not in leaves but in potato tubers and that an oral vaccine may then be available.

A team of investigators from the Pasteur Institute of Iran (Tehran) took up this challenge (Domansky *et al.*, 1995; Ehsani *et al.*, 1997). In their first (brief) publication, these authors (Domansky *et al.*, 1995) followed the procedure of Mason *et al.* (1992) but used a different source for the coding sequence of HBsAg subtype: *ayw,* that was obtained from K.G. Gazaryan (Russian Academy of Sciences) and transformed potato plants. But this publication deals only with leaves and roots of the transgenic potato. Tubers were not analyzed. The level of HBsAg found in the potato leaves was similar to those found by Mason *et al.* (1992) in tobacco leaves. In a further publication of this group (Ehsani *et al.*, 1997), it was mainly intended to produce the HBV M protein and to compare it with the production of the HBV S protein. It should be noted that there are three envelope proteins of HBV: the large (L), the middle (M) and the small (S). All three are encoded in one large open reading frame. The resulting polyprotein is then processed into pre-S1, pre-S2 and S. The pre-S2 yields the M protein. The M protein is considered highly immunogenic and therefore useful in HBV vaccination. The investigators thus constructed two vectors. In one, the gene for the S protein was used while in the other vector, they included a DNA sequence that encoded the pre-S2 protein (to yield the M protein). After transformation with the sequence coding for the S-protein and due purification, they found multimeric (mostly trimeric) forms of HBsAg in the potato tissues. The plants transformed with the sequence that encoded the pre-S2 yielded an antigen that was apparently highly glycosylated in the plant tissue. The quantities of HBsAg in roots, leaves and tubers were estimated respectively as 80, 10 and 60 ng/mg soluble protein. This is still a rather low level. Unfortunately, no further reports on the antigens of HBV appeared in reviewed literature. Was further work done in commercial laboratories or did the investigators abandon this line of research? We have no way to know.

A collaborative effort by ten investigators from Warsaw, Poznan and Philadelphia (Kapusta *et al.*, 1999) brought the plant-derived oral vaccination against HBV one step further. These investigators transformed lupin (*Lupinus luteus*) and lettuce (*Lactuca sativa*) and obtained the respective transgenic callus and transgenic leaves that expressed the envelope surface protein of the hepatitis B virus. The levels of this antigen reached up to 150 ng or 5.5 ng per g fresh weight in the callus or in the leaves, respectively. Three human volunteers were fed twice with the transgenic lettuce leaves (200 g and after two months another 150 g). After the second feeding, two of the volunteers produced a level of HBsAg-specific antibody that was above the minimal level for protection against HBV, but this level declined to nil after 12 weeks. No such antibody was found in the two volunteers who were fed control lettuce leaves. We noted that the total number of volunteers was only half the number of authors in this publication.

4.2.1.2. *Rabies virus*

We shall move from a pathogenic virus that has a very limited host range (man and some higher primates) to a pathogenic virus with a vast host range (man, dog, vampire bats and many other mammals, and is transmitted through bites of rabid animals). In other words: from HBV to rabies virus. The latter was the one handled by Louis Pasteur, as mentioned in Chap. 3. This is a deadly disease and passive immunization will rescue infected people (when applied in time) but it is problematic. It can be given once, but if a second dose of antibody is given (with serum that is not of human source), death may be caused by the alien serum that comes with the antibodies. The problem is especially severe in areas where contaminated vampire bats are common (e.g. in some South American countries) as well as in developing countries or countries that border with developing countries, where infected animals of the canine family are roaming. A large team of investigators from Philadelphia and USDA, Beltsville, Maryland, intended to produce, ultimately, an edible vaccine against rabies. This vaccine should be inexpensive so that mass immunization of humans against rabies can be practiced in developing countries. As a first step towards this goal, this team (McGravey *et al.*, 1995) explored the possibility to express the rabies G-protein (the only immunogenic protein of rabies) in tomato. These investigators used a standard procedure to transform tomato plans by *Agrobacterium tumefaciens* (the binary vector procedure). Their binary vector (RG-2) contained the *npt*II gene for kanamycin resistance, flanked by the *Nos* promoter and the *Nos* terminator. Tail-to-tail with this cassette, the

RG-2 contained the complete, unmodified G-protein gene from the ERA strain of rabies virus, and it was flanked by the CaMV 35S promoter and the *Nos* terminator. Hence, no effort was invested to express RG-2 specifically in the fruits; neither was there an effort to retain the G-protein in the ER nor to secrete it into the apoplast. After transformation and selection on kanamycin-containing medium, they retained four transgenic plants, two of these contained the transgene. Only one of these was selected for self-pollination (Rgp-13) and it resulted in 22 plants, 18 of which contained the alien gene. The authors concluded that this indicated that the transgene was integrated only once in the tomato genome (the statistics would also fit two integrations). Northern blot analysis showed hybridization of total RNA from either the leaves or fruits of Rgp-13 with the rabies G-protein gene. In Western blot analysis, the protein extracted from the fruits and leaves of Rgp-13 gave two bands (apparent molecular weights ~ 62.5 and 60 kDa), while the protein from authentic rabies G-protein showed only one band (~ 66 kDa) when probed with mouse monoclonal anti-G-protein antibodies. The authors attributed the difference in migration to different glycosylations in mammals versus those in plants. The estimated levels of G-protein in transgenic tomato plants were disappointing: 1–10 ng per mg soluble protein in the leaves and even less in the fruits. On the brighter side of this report was the observation that the G-protein could be revealed by immunogold labeling of leaf tissue. What happened with this system in subsequent years is enigmatic. We saw no further reports in reviewed journals on the production of rabies G-proteins by transgenic plants.

4.2.1.3. *Norwalk virus*

Norwalk virus (NV) causes acute epidemic gastroenteritis in human populations. NV is a member of the family Caliciviridae which is spread by ingestion of contaminated food or water and by secondary person-to-person transmission. There is an experimental problem with this virus. Up to now, neither the NV nor related viruses can be grown in cell cultures (see further details on NV in Ball *et al.*, 1999). NV is a medically important virus because it is a major cause of acute epidemic gastroenteritis (over 40% in the USA).

The Mason–Arntzen team (Mason *et al.*, 1996) directed their attention to NV with the intent of developing an oral vaccination against NV that will be produced by transgenic plants. The NV capsid protein (NVCP) was previously expressed in insect cell cultures. It is composed of 90 homodimers (the Mr of each monomer is 58,000), and attains a diameter of 38 nm. The empty capsids are morphologically similar to the NV but lack the viral genome. The NVCP is immunogenic: 50 μg of the insect-cell-derived NVCP (I-rNV) applied orally to a mouse will elicit specific serum and mucosal antibodies. Moreover, a recent study (Ball *et al.*, 1999) showed that the I-rNV particles can be safely given to human volunteers (at a dose of 250 μg) and will elicit IgG and IgA responses without causing deleterious side effects.

These investigators (Mason *et al.*, 1996) constructed three plasmids as vectors for *Agrobacterium*-mediated plant transformation. All these three plasmids had the T-DNA right and left border sequences. They also contained the *npt*II gene for kanamycin resistance (flanked by an appropriate promoter and a terminator). In pNV101, the coding sequence for NVCP was inserted downstream of the 35S CaMV promoter with a double translational enhancer. In pNV102, they also added to the 3′ of the 35S promoter a duplicated enhancer region but with the addition of the untranslated TEV region. The pNV140 had the patatin promoter, followed by the TEV region, in front of the coding sequence for NVCP. All the three plasmids contained the *Nos* terminator, downstream of the coding sequence for NVCP. When tobacco leave sections were transformed with agrobacteria containing either pNV101 or pNV102, transgenic plants that produced various levels of mRNA for NVCP were obtained.

The TEV 5′ UTR had a beneficial effect on translation: transgenic tobacco plants transformed with pNV102 produced threefold more NVCP antigen than those transformed with pNV101. The maximum accumulation of NVCP in transgenic tobacco leaves was 0.23% of the total soluble protein. Similar levels of NVCP were estimated in the tubers of potato plants transformed with pNV140. Several tests indicated that the tobacco-produced NVCP (t-rNV) is very similar to the NVCP produced in insect cell cultures (i-rNV). For example, the i-rNV and the t-rNV particles had exactly the same shape as observed by negative staining and electron microscopy (Fig. 13). The oral immunogenicity of the t-rNV and the potato-tuber rNV was also assessed by direct gastric application (garage) to CD1 mice. The mice received 10–80 μg doses of t-rNV on days 1, 2, 11 and 28. Control mice received the equivalent amounts of leaf extract from nontransformed plants. Other mice were fed with 4 g of tuber tissue (estimated as 40–80 mg of rNV) also on days 1, 2, 11 and 28. In some tests, cholera toxin (CT) was added to the rNV. Fecal samples were

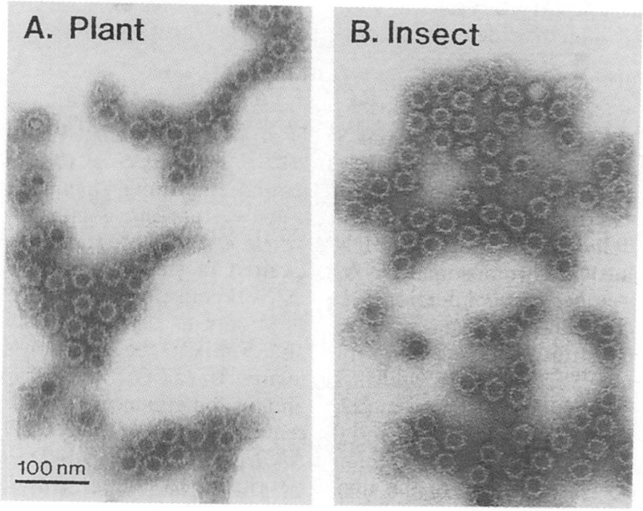

Fig. 13. I-rNV and t-rNV particles visualized by negative staining and electron microscopy. The peak fractions from ion exchange chromatography of t-rNV(A) or i-rNV(B) were pooled and pelleted at 10,000 × *g* and processed for negative staining and electron microscopy. Bar = 100 nm. (From Mason *et al.*, 1996.)

collected and tested for total IgA and anti-NVCP IgA. The addition of CT had a positive effect on antibody production when t-rNV was given by gastric intubation but there was no CT effect when the potato-tuber rNV was fed to mice. Specific IgA levels reached a maximum of 4 ng/μg after treatment with t-rNV. The serum titer of mice fed with the potato-tuber rNV was lower than the respective titer in mice treated with t-rNV. The authors concluded that their study provided "further evidence to support the concept of edible vaccines". On the other hand, they admitted that neutralization activity against NV was not yet shown, and they intended, in the future, to conduct human feeding trials for direct assessment of protection afforded by ingestion of recombinant potato tubers against Norwalk virus infection.

Several studies were devoted to explore the production by plants of immunogenic proteins that will elicit antibodies against animal (nonhuman) viral diseases.

4.2.1.4. *Swine-transmissible gastroenteritis virus*

This virus (TGEV) is the causative agent of acute diarrhea of newborn piglets. TGEV provokes high mortality and is therefore of great economical importance in the pork industry. The piglets may be protected if the respective antibodies in the pregnant sows are raised to a sufficient high level. The protection is then transferred through the placenta, the colostrum and milk. The neutralizing antibodies against TGEV are directed mainly to the envelope glycoproteins.

The relevant epitopes for this neutralization have been mapped in the N-terminal domain of this glycoprotein S (spike). Four major antigenic sites were described in this glycoprotein, of which site A is the immunodominant (see Gomez *et al.*, 1998 for further information and references). Gomez *et al.* (1998) investigated the feasibility of expressing the glycoprotein S of TGEV in transgenic plants and to explore the immunogenicity of the plant-derived protein. The glycoproteins of TGEV are resistant to degradation in the swine gut and therefore can serve as a good model for oral vaccines. These authors choose the model plant *Arabidopsis thaliana* rather than a crop plant to express the antigen. They constructed two (*Agrobacterium*) binary plasmids. These plasmids had the same components but for the transgene that should lead to the expression of the antigen. The plasmids had a kanamycin-resistance cassette and the other cassette (orientation: tail-to-head) was for the expression of the transgene. This cassette contained the 35S CaMV promoter and the *Nos* terminator. The transgene in plasmid pRoK I was the code for the N-terminal domain of the glycoprotein S from TGEV and the transgene in pRoK II was the full-length glycoprotein S of this virus. The tranformation of *A. thaliana* was performed as described by us in Sec. 2.4.1.2. The inflorescences of the plants were dipped in a suspension of *Agrobacterium* containing either pRoK I or pRoK II, and the bacteria were infiltrated by partial vacuum. Due selection on kanamycin-containing medium was performed and analysis of the plants was performed on the sexual progeny of transformed plants that were self-pollinated (the authors termed this generation "F_2", which is not the term that would appeal to geneticists). Twenty such progeny lines, for each plasmid, were found to harbor the transgene. Four plants from each of these 20 lines were used for further analyses. All produced the respective transcripts (i.e. for either the full-length glycoprotein or for the N-terminal domain). But when Western blot hybridization was performed, none of the transgenic plants was found to produce the glycoprotein S or its N-terminal domain. The authors assumed that the proteins were produced but at levels that were too low for detection (they estimated 0.03 to 0.06% of glycoprotein S in the total soluble protein in the *Arabidopsis* leaves). Nevertheless, BALB/c mice were immunized (one mouse for each transgenic plant) intramuscularly (days 0, 15 and 30) with 30 μg total leaf protein (in PBS and complete Freund's adjuvant in the first immunization). The sera were then tested (ELISA) with purified TGEV as antigen. All sera showed that they had antibodies against TGEV. Virus neutralization was also observed after reaction with the sera of mice immunized with the extracts of the transgenic plants. The authors admitted that the level of antigen production by the transgenic plants should be increased. They believed that the use of different promoters, plant-derived leader sequences and signal peptides as well as modification of the codons should lead to higher expression in the transgenic plants.

4.2.1.5. *Human cytomegalovirus (HCMV)*

HCMV is an important human disease that is especially severe or even lethal in low-weight preterm infants and immunocompromised patients such as patients after organ transplantation and AIDS patients. The cDNA for the immunodominant glycoprotein B complex (gB) is known and gB is an antigen for HCMV. Takaberry *et al.* (1999) constructed a plasmid for *Agrobacterium*-mediated transformation that should express gB in seed protein. The plasmid contained sequences of the rice glutelin gene to direct the expression into the seeds. The genetic transformation resulted in 28 transgenic tobacco plants; in most of them, there was the gene for gB. The seeds of the latter plants produced gB protein in the range of about 70 to 150 ng per mg total protein. There was similarity between the tobacco-seed-derived gB and natural gB. In this respect, the study constituted an important contribution to plant-derived antigen production because the natural formation of gB is involved in a complicated process of cleavages, folding and glycosylations.

4.2.1.6. *Foot-and-mouth disease virus*

Another Spanish/Argentine collaboration (Carrillo *et al.*, 1998; Wigdorovitz *et al.*, 1999) studied the possibility of producing antigens for an additional virus: the foot-and-mouth disease virus (FMDV) that affects ruminants and, also occasionally, man.

FMDV is a very important disease of domestic ruminants. The disease spreads easily across borders by wild ruminants and the safest remedy is by mass vaccination of all domestic ruminants in a confined area. There are vaccines in use that are based on inactivated virus and these vaccines are effective. But, these vaccines are expensive and carry a risk: the producing laboratories have to be under strict surveillance, to avoid release of live viruses into the environment. On the other hand, all potentially sensitive hosts (domesticated and wild) in the confined area should be vaccinated. A structural protein VP1 carries critical epitopes that can induce the production of neutralizing antibodies in animals vaccinated with VP1. Such an induction was also reported after immunization with synthetic sequences of amino acids which represent specific parts of VP1.

The intention of these investigators was to express VP1 in transgenic plants. They chose *Arabidopsis thaliana* for *Agrobacterium*-mediated transformation and used a very similar system to that used in these laboratories in a previous study to produce TGEV vaccines by *A. thaliana* (Gomez *et al.*, 1998, see above). The gene for VP1 was amplified by reverse-transcription-PCR from viral RNA of FMDV and inserted in an appropriate binary plasmid. The transformation was performed by dipping (with vacuum) the inflorescence of *A. thaliana* plants into a suspension of *A. tumefaciens* that contained the binary plasmid (which also contained the *npt*II selectable gene). After self-pollination of the plants that germinated from seeds produced in the transformed inflorescence, the plants that were resistant to kanamycin were used for further analysis. The authors then used PCR to detect a 145-bp fragment, which codes for VP1, in the analyzed plants and detected several plants that contained this fragment. Further tests were performed by ELISA and 1/10 of the plants were found to express a protein that was bound specifically to FMDV antibody. Western blot hybridization with anti-FMDV antibodies confirmed the production of VP1 in some of the transgenic *A. thaliana* plants. Extracts of plants that apparently expressed the VP1 were injected (three times) intraperitoneally to male (BALB/c) mice. The sera of these mice were found to contain anti-VP1 antibody and the sera showed immune response against intact FMDV particles. Finally, mice immunized with extract of the transgenic plants showed resistance against challenge with virulent FMDV. Hence, in this study, the research reached a further step: a plant-produced vaccine was shown to protect animals against a viral challenge. Only the last sentence in this publication is surprising: "These findings support the concept of using transgenic plants, as a novel and safe system of inexpensive vaccine production, which could become *a very attractive alternative in the developing world*." Well, the term "developing world" is a term that makes sense only by its political correctness. The authors probably mean areas of poor populations. But the main enigma is: if production of vaccines by plants is safe and efficient, why not employ it throughout the *whole world*?

In a following publication, this team (Wigdorovitz *et al.*, 1999) tested the shift from intraperitoneal injection of plant extract to oral immunization and from *A. thaliana* to alfalfa (*Medicago*). The latter is a regular fodder for ruminants. The first phase of this study was very similar to that of the previous (Carrillo *et al.*, 1998) study. The only difference was that alfalfa plants rather than *A. thaliana* were transformed. This required the use of the regular *Agrobacterium*-mediated transformation with due modifications to adapt the protocol to alfalfa. Briefly, the coding sequence for VP1 was detected in transgenic alfalfa plants by PCR. Then, 15 transgenic plants that contained the VP1 gene were analyzed for the recombinant VP1 protein by ELISA and 10 out of these plants reacted with anti-FMDV serum. The authors first repeated the immunization performed previously with *A. thaliana* plant extract. In the present study, they used alfalfa plant extract to immunize mice. It was found that the immunized mice produced the expected antibody. Then, the oral immunization was tested by feeding mice with 0.3 g of fresh leaves, three times a week, for two weeks. The anti-FMDV antibodies were found in the blood of immunized mice, ten days after the last feeding of the transgenic alfalfa leaves. The mice that were orally fed with transgenic leaves developed a specific antibody response against whole virus particles, while control mice fed with regular alfalfa leaves showed no such response. Finally, 77 to 80% of the mice that were immunized intraperitoneally with extract from the transgenic alfalfa were protected against a challenge of FMDV, and 66 to 75% of the mice that were immunized orally were also protected against this challenge. It is not clear on how many mice these percentages were based. There is still one further step missing: using ruminants rather than mice to evaluate the capacity of oral immunization by a plant-derived antigen.

4.2.1.7. *Rabbit hemorrhagic disease virus*

Recently, Castanon *et al.* (1999) explored the immunization by transgenic potato plants against a rabbit disease. Rabbit hemorrhagic disease (RHD) is caused by a virus of the Caliciviridae family. It is a deadly disease. Infection of adult rabbits causes an acute liver damage and disseminated intravascular coagulation. The disease is controlled in domestic and commercial rabbit populations by slaughter and vaccination. This rabbit/RHDV system has several interesting features. First, the VP60, which is the major structural protein of RHDV, was produced previously in several heterologous systems and it was found to induce protection in rabbits against a lethal challenge of RHDV. The coding sequence of VP60 was available and could be used in a plant transformation vector. Finally, the system has the advantage that the same animals (rabbits) which are the natural hosts of RHDV are small enough to be used as "model" animals (rather than the use of mice in previous studies, mentioned above).

Castanon *et al.* (1999) thus inserted the coding sequence for VP60 into plant vectors for *Agrobacterium*-mediated transformation. The vectors also contained cassettes for *npt*II (kanamycin resistance). In one vector (pK2-VP60), they inserted a regular 35S CaMV promoter upstream of the coding sequence for VP60 while in another vector (pK3-VP60), the 35S CaMV promoter was followed by a double translational enhancer sequence. Potato leaves were transformed with *Agrobacterium* suspensions that harbored either pK2-VP60 or pK3-VP60. After due selection (and regeneration), they obtained putatively transgenic potato plants. All these plants contained the coding sequence for VP60 (as revealed by PCR tests). Northern blot analyses showed that all nine putative transgenic potato plants transformed with pK2-VP60 had the VP60 transcript and seven out of the nine putative transgenic potato plants transformed with pK3-VP60 produced this transcript. No transcript of VP60 was found in control potato plants. For detection and quantification of recombinant VP60 in leaf extracts, the investigators performed a sandwich ELISA test in which they used the respective monoclonal antibody against VP60. This test showed that the leaves of plants transformed with pK2-VP60 had rather low levels of VP60 while in leaves of plants transformed with pK3-VP60, these levels were much higher (0.8 to 3 μg per mg of soluble protein). This means that the double translational enhancer between the 35S CaMV promoter and the sequence coding for VP60 substantially increased the production of VP60 in the leaves of the transgenic potato plants. The size of the VP60 protein in the transgenic plants was the same as the size of native VP60, having a molecular mass of 60 kDa. But, no virus-like particles could be observed in the tissue of transgenic plants. The authors assumed that this was due to the very small amount of VP60 produced in the plants. Two rabbits were inoculated four times (days 0, 30, 60, 90) with 1 ml of leaf extract that contained 12 μg of VP60 emulsified in complete Freund's adjuvant. The first immunization was by the subcutaneous route (in several points) whereas the boosters were given intramuscularly. As negative control, two rabbits were injected with extracts of wildtype potato plants. High antibody titers were found in the rabbits injected with extracts of the transgenic plant. Thereafter, the rabbits were challenged with virulent RHDV (16,000 hemagglutination units per rabbit). The two control rabbits died 40 hours after challenge while the two rabbits which received the plant extract with VP60 were fully protected. Well, while "one swallow does not bring the spring", these authors claimed that two rabbits proved their point: rabbits can be protected against RHDV by tissues of transgenic plants that produce VP60. On the other hand, the authors admit that, for oral immunization (which is a legitimate goal), the VP60 content in tubers of transgenic potato should be substantially higher.

Finally, the authors had an interesting reason why a cheap oral vaccination against RHDV is important. This reason is not based on benefiting a "developing" country but the authors state: "... using edible plants for oral immunization is an *important issue* considering that RHDV also infects wild rabbits which constitute an important staple prey for some protected carnivores (e.g. *Linx pardinus* — should be *Lynx pardinus)* and is the most important small game species in Spain". Should we expect future contributions from the Spanish Hunter Association to develop such an oral vaccine? If we follow the logic of the eight authors, one could say that an effective and cheap oral immunization against RHDV will save the wild rabbits of Spain, satisfy the lynx population and thus provide enjoyment for the Spanish hunters.

4.2.2. *Antigens from bacteria*

4.2.2.1. *Vibrio cholerae and the enterotoxigenic Escherichia coli*

The Gram-negative bacterium *Vibrio cholerae* is the pathogen that causes the watery diarrhea called cholera. This disease is known from ancient times, recorded at least 2,000 years ago. Paradoxically, the disease should theoretically hardly occur and when it does occur, it can be cured easily; but in practice, it is frequently lethal to great masses of people, especially children. This disease caused recurrent pandemics since about 130 years ago in the "developing" countries. The infection occurs through the digestive tract (i.e. the small intestine) and when an epidemic occurs, in areas with bad sanitation, drinking water and food recycle the bacteria lavishly. When an infected person is diagnosed, rehydration and other medical treatment can cure the person within a few days but when cholera occurs in a dense population, thousands are suddenly infected, the medical facilities are just not sufficient to treat all patients — those not treated have a high rate of mortality. The result: in our time, with the advanced medical care in the "developed" world, there are still about five million cases of cholera annually in the "developing" world and about 200,000 people die as a result of it. There are several biotypes of cholera. A very virulent biotype was identified during an epidemic in Peru (in 1991). The epidemic strains also differ in the serotype lipopolysaccharide antigen. A relatively recent one is the serotype O 139 or "Bengal" serogroup of *V. cholerae.* Thus, vaccination against cholera is problematic due to the new emerging serogroups (see Mekalanos and Sadaff, 1994, for review). Resistance to cholera requires a strong *mucosal* immunity and the immunity should be long lasting. In the past, vaccinations gave only a short (about three months) and partial protection. Lately, the vaccination became more successful. One method was to vaccinate with a mixture of the nontoxic B subunit of the cholera toxin and killed whole bacteria (BS-WC). In some trials, this vaccination provided 85% protection (in about 63,500 individuals). Such a protection, if it lasts for a sufficient long period, is actually solving the problem because if the number of affected people is reduced from 10,000 to 1,500, there is a chance that the latter can be saved by treatment in hospitals. Another vaccination (still on trials) is the use of live attenuated *V. cholerae* vaccines provided orally in a single dose. Several modifications of the above mentioned vaccinations are on trials, some based on mutants of the pathogenic *V. cholerae* bacterium, and the aim is to vaccinate the whole population in endemic areas. Obviously, in parallel, improving the sanitary conditions will have a major impact and reduce the outbreak of cholera epidemics. The enterotoxigenic *Escherichia coli* bacterium (ETEC) also causes acute watery diarrhea. Both bacteria colonize the small intestine and produce one or more enterotoxins. These include the heat-labile enterotoxin (LT) of ETEC. The LT of ETEC has an identical structure to that of the cholera toxin. It has one large subunit (A, 27 kDa) that is toxic and a pentamer of small binding subunits (B, 11.6 kDa). The latter was termed LT-B. The toxin binds to the G_{m1} ganglioside in the epithelial cell surface of the intestine. An antibody (mucosal) that will interfere with the binding of the B subunits with the epithelial cell surface should block the toxic activity. The heat-labile enterotoxin (LT) of ETEC and the cholera enterotoxin (CT) have the same basic structure and have similar functions and immunochemistry. We shall therefore handle them in the same section.

The team of Arntzen and collaborators, who pioneered in the attempt to produce a plant-derived vaccine against HBV, turned to the production of such an oral vaccine against ETEC (Haq *et al.*, 1995). These investigators used the LT-B of the pathogenic *E. coli* to produce a plant-derived oral vaccine. They engineered two expression vectors (binary plasmids) for *Agrobacterium*-mediated transformation of tobacco and potato plants. In one vector (pLT-B-110), they inserted the TEV untranslated leader between the cDNA for LT-B and the 35S CaMV and a terminator from the soybean *vsp*B gene. The vector also included a kanamycin-resistance (*npt*II) cassette and had the LB and the RB of the T-DNA of *Agrobacterium tumefaciens*. The second vector (pLTK-110) was similar to pLT-B-110 but the cDNA for LT-B was followed by the code for ER retention: amino acids Ser, Glu, Lys, Asp, Glu, Leu (SEKDEL). Both pLT-B-110 and pLTK-110 were used to transform tobacco and potato by the conventional procedures. Leaves from putative tobacco transformants and microtubers of putative transgenic potato (microtubers were produced according to the procedure developed by one of us, Perl *et al.*, 1992) were then analyzed. The amount of recombinant LT-B was much higher in the microtubers than in the tobacco leaves. The maximum levels of LT-B in microtubers were in potato plants transformed with pLTK-110: 110 mg/g soluble protein (less than 15 mg/g soluble LTB protein was in tobacco leaves after transformation with pLTk-110 and only up to 5 mg/g after transformation of tobacco plants with pLT-B-110). Hence, microtubers were a better source of recombinant LT-B than leaves and the ER retention signal vastly improved the accumulation of recombinant LT-B. Analysis by electrophoresis indicated that the plant-produced rLT-B is similar to the authentic LT-B and the rLT-B from plants that were transformed with the SEKDEL signal showed structural and binding characteristics that were similar to those of rLT-B produced in *E. coli.* Thus, for oral immunization, the rLT-B from the latter plants was utilized. Each oral dose contained 12.5 μg of antigen (rLT-B) and the vaccine was administered to mice on days 0, 4, 21 and 25. Another group of mice received a similar treatment of *E. coli* expressed LT-B. Serum antibodies and mucosal antibodies were then evaluated. The neutralization of *E. coli* enterotoxin activity by serum and mucosal extracts from the mice that were orally immunized with the tobacco leaf extract was effective and similar to the results with vaccines based on *E. coli* produced LT-B. Mice (BALB/c) were also orally immunized by feeding them slices of microtubers from transgenic potato plants. As for oral immunization with transgenic microtubers, each mouse received 5 g (for two to six hours) of microtuber tissue that was equivalent to 15–20 μg of rLT-B (SEKDEL). The mice developed serum IgG and mucosal IgA that were specific to LT-B, but the levels were lower than those in mice fed with rLT-B from *E. coli.*

In a further publication of the same group (Mason *et al.*, 1998), the work on a plant-derived antigen against the LT of *E. coli* was brought one step further (but in this later report, the ER SEKDEL signal was ignored!). The coding sequence of the bacterial cDNA for LT-B was changed to better fit the plant codon usage. Thus, a synthetic gene sequence was produced (sLT-B). To enable easier engineering of the expression vector, they also changed the second codon from AAT to GTG (i.e. from coding Asn to Val). Otherwise, the new vector was identical to the pLT-B-110 mentioned above (i.e. without the code for SEKDEL). After the appropriate transformation, 75 putative transgenic potato plants were obtained. Of these, 20 plants had (evaluated by ELISA) various levels of LT-B. The highest line (TH-110-51) contained LT-B at 1.9% of the total soluble protein in the leaves, when grown in the growth chamber. When transferred to soil (in a greenhouse), the level of LT-B was reduced. Northern blot analysis showed that the line with the highest LT-B also had the highest sLT-B transcript. The transgenic potato plants that produce LT-B were grown further to produce tubers. The skin of the tubers was removed and they were used for feeding mice. Tubers of several potato lines were used and mice were also fed with bacterial LT-B. The tubers of TH-110-51 contained 10 μg/g of LT-B. Mice were fed 5g-tuber doses on each of the days: 0, 7 and 14. Indeed, the feeding of the transgenic potato tubers, such as of the line TH-110-51, elicited in the mice the production of anti-LT-B fecal IgA and anti-LT-B serum IgG. No such responses were found after feeding mice with control potato tubers.

Protection against toxin challenge was tested in potato-fed mice by a "patent mouse assay". The LT (as well as CT) elicits fluid secretion into the bowel and the fluid accumulates in the intestine. Treated mice are thus sacrificed and the ratio of gut/carcass is evaluated. It was found that for mice which were first fed with either purified bacterial LT-B or slices of transgenic potato tubers and were subsequently challenged with LT, there was a reduction of the gut/carcass ratio. But, neither the bacterial LT-B nor the transgenic tuber feeding caused a complete protection against the LT challenge.

The Mason–Arntzen team, now in collaboration with investigators from the Department of Medicine, University of Maryland School of Medicine (Baltimore, MD, USA) brought the ETEC system to the stage of preclinical studies (Tacket *et al.*, 1998). They used potato tubers of the transgenic line TH-110-51 (mentioned above). Human volunteers were given either 100 g of transgenic potato tubers, 50 g of the same transgenic potato tubers or 50 g of "wildtype" potato tubers. The content of LT-B in the potato tissue varied considerably (~ 4 to 16 μg/g). The authors estimated that the 50 or 100 g potato contained in total approximately 0.4 or 1.1 mg of LT-B ("mean 0.75 mg/dose"). The volunteers were given three doses of potato: on days 0, 7 and 21. The potatoes were raw but the 50 or 100 g were generally well tolerated by the volunteers, only two of them who ate 100g doses of raw potato complained of some nausea.

The efficiency of the potato-derived antigen was evaluated by testing serum and mucosal immune responses in the volunteers. One test was based on gut-derived antibody-secreting cells (ASC). ASC are cells that secrete specific antibodies. They commonly appear one week after mucosal immunization. It was found that, in volunteers who consumed the transgenic potatoes, there was an increase, seven days after the first dose of potatoes, of anti-LT IgA ASC per 10^6 peripheral blood mononuclear cells (PBMC). This higher level was maintained for four weeks. In 10 out of 11 volunteers who ate the transgenic potatoes there was an increase in anti-LT IgG. This increase was not detected in volunteers who ate "wildtype" potatoes — but the increase was not very impressive. Stools were collected and assayed for anti-LT SIgA. In five out of ten volunteers who ate transgenic potatoes, there was a fourfold rise of this SIgA. The authors claimed that "these results offer a new strategy for developing safe and inexpensive vaccines against diseases for which a protective antigen has been defined, such as tetanus, diphtheria and hepatitis B". They suggested transgenic banana for future oral immunization. One cannot argue on this suggestion... to most people, bananas are preferred over raw potato tubers.

A study on the production of an oral vaccine against cholera, based on an antigen produced by transgenic plants, was performed by another team of researchers (Arakawa *et al.*, 1997; 1998a) who actually submitted their papers before the second paper of the Arntzen–Mason group (i.e. Mason *et al.*, 1998). The former group had the same strategy as the Arntzen team, meaning that the production of the oral vaccine should be by transgenic potato and the expression vector should encode the cholera CTB (which is similar to the LT-B of ETEC). In both cholera and ETEC, the toxic subunit of the enterotoxin is the A subunit but the pentamers of B subunits cause binding of the enterotoxin to the G_{m1} ganglioside. If antibodies interfere with the binding of the B subunit — the toxic A subunit will not affect the intestine cells. The authors (Arakawa *et al.*, 1997; 1998a) added the SEKDEL signal (for retention in the ER) downstream of the coding sequence for CTB.

The authors constructed an interesting expression vector for *Agrobacterium*-mediated genetic transformation of potato (Fig. 14). The CTB coding sequence (and the preceding leader sequence) was driven by the *mas* P1.2 promoter. This is the bidirectional and auxin-activated mannopine synthase promoter of *A. tumefaciens*. Thus, this promoter

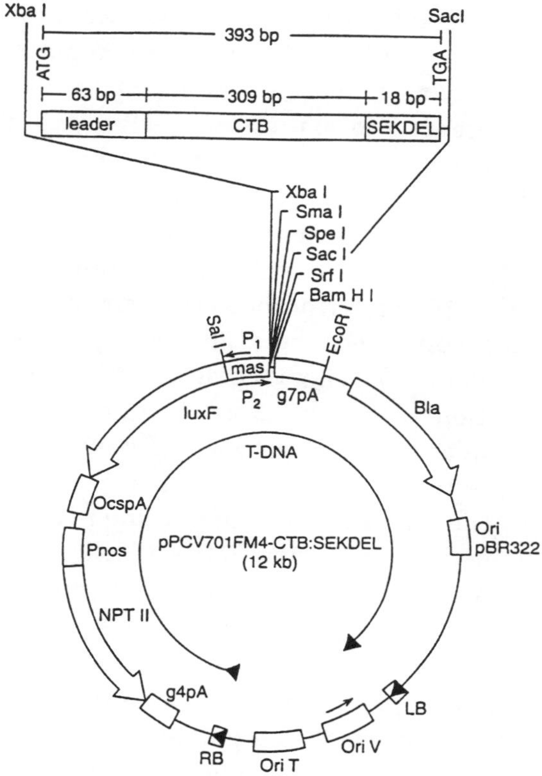

Fig. 14. Structure of the plant expression vector pPCV701FM4-CTB-SEKDEL. The following four genes are located within the T-DNA sequence flanked by the right and left border (RB and LB) 25-bp direct repeats required for integration of the T-DNA into the plant genomic DNA: (1) a 393-bp CTB: SEKDEL coding sequence under control of the *mas* P2 promoter; (2) the bacterial luciferase AB fusion gene (*lux*F) under control of the *mas* P1 promoter as a detectable marker; (3) an NPT-II expression cassette for resistance to kanamycin in plants; (4) a β-lactamase cassette for resistance to ampicillin in *E. coli* and carbenicillin in *A tumefaciens*. The g7pA polyadenylation signal is from the *A. tumefaciens* T_L-DNA gene 7; the OcspA polyadenylation signal is from the octopine synthase gene; Pnos is the promoter of the nopaline synthase gene; g4pA is the polyadenylation signal from T_L-DNA gene 4; Ori T is the origin of transfer derived from pRK2; Ori V is the wide host range origin of replication for multiplication of the plasmid in *A. tumefaciens* derived from pRK2; and Ori pBR322 is the replication origin of pBR322 for maintenance of the plasmid in *E. coli*. (From Arakawa *et al.*, 1997.)

activated, in one direction, the CTB coding sequence and, in the other direction, it activated the bacterial luciferase reporter gene (*lux*F). This enabled the authors to analyze (even *in vivo*) the expression of the chimeral transgene. The expression vector also included other regulatory components such as a cassette for kanamycin resistance and the RB and LB of the T-DNA of *A. tumefaciens*. In addition, the code for a leader peptide (21 amino acids) upstream of the coding sequence for CTB was also included. Among the putatively transformed potato plants, two showed a high luciferase activity. These plants also contained the CTB coding sequence (by PCR detection). The maximum level of CTB detected in leaves of transgenic plants after induction with auxin was 35 μg/g fresh weight. Much lower levels were detected in microtubers of transgenic potato plants. The potato-produced CTB was found to have a strong affinity to the G_{m1} ganglioside. The authors found that in the plant tissue, the CTB tended to dissociate into monomers (M_r ~ 15 kDa) but plant-derived CTB did have immunological and biochemical properties that were as those of the native (bacterial) CTB.

In a further study, these authors (Arakawa *et al.*, 1998a) used the same transgenic potato plants to bring the system closer to medical application. They fed mice with the potato tissue, four times, for one month and found specific CTB antibodies in the feces and sera of these mice. Anti-CTB titers for IgM and IgA were slightly increased from day 35 to day 70 but the anti-CTB for IgM decreased subsequently. The authors also performed a cholera toxin neutralization assay (Vero cells in culture) with antisera of mice fed with the transgenic potato plants. This antiserum neutralized the CT effect. When mice were first vaccinated with potato tissue and were then challenged with CT, the mice showed a 60% reduction in diarrheal fluid accumulation. The vaccinated mice also showed that there was a reduction of the CT binding to the cell surface receptor G_{m1} ganglioside. While the studies of the Arntzen–Mason group and of Arakowa and associates give hope for an edible plant-derived vaccine against cholera, the road to its medical application is still long. There are still several improvements in the expression vector that can be performed and uncooked potato tissue may not be the best source for a human edible vaccine.

4.2.3. *Protection against autoimmunity*

Insulin-dependent diabetes mellitus (IDDM) is an autoimmune disease that afflicts about 0.25% of the human population. It is caused by the autoimmune destruction of insulin-secreting pancreatic β-cells.

Autoimmune diseases can be treated in some specific cases by oral administration of a protein that will cause tolerance to autoimmunity. But inducing oral tolerance to autoimmunogens requires large amounts of the respective proteins. Recent studies, reported below, suggested that edible transgenic plants may be a good source for the required proteins. Ma *et al.* (1997) tested this approach in a mouse model for diabetes studies. The enzyme glutamic acid decarboxylase (GAD) has been implicated as an autoantigen in diabetes (see Yoon *et al.*, 1999). Anti-GAD antibodies appear before the onset of diabetes in human patients and in non-obese diabetes mice (NOD). The NOD mice are a model of spontaneous insulin-dependent diabetes mellitus. It was previously reported that immunization of young NOD mice with GAD suppressed diabetes. No oral supply of GAD was tested previously in NOD mice probably due to the requirement of very large GAD quantities. Ma *et al.* (1997) went to test oral application to NOD mice and to furnish GAD produced in transgenic plants. They focused on GAD67 which is the isoform predominating in mouse islets. They obtained the cDNA of GAD67 and inserted it into an *A. tumefaciens* expression vector. This vector had the usual components but the TEV untranslated leader was inserted downstream of the 35S CaMV promoter with a double enhancer sequence and the initiation site of the cDNA for GAD was modified for the easier engineering of the vector. The expression vector was based on the pBin19 plasmid and was called pSM215. The conventional *Agrobacterium*-mediated genetic transformation was performed to produce transgenic tobacco (a cultivar with low alkaloids) or transgenic potato plants. Transgenic tobacco plants and potato plants were obtained and found to contain about 0.4% of GAD in their total soluble protein. The mice were fed daily with plant material that was estimated to contain 1–1.5 mg of GAD. A four-week course of such feeding of NOD mice seemed to inhibit the development of diabetes. Such GAD-containing plant material, when fed to NOD mice from five weeks to eight months (daily) prevented diabetes in 10 out of 12 mice. In control feeding (plant material that did not contain GAD), 8 out of 12 NOD mice did develop diabetes. The authors considered that, in the future, the human GAD65 can be expressed in human food plants and thus can be used in preclinical tests.

Indeed, this challenge was taken up by 11 Italian investigators (from Perugia, Siena and Verona). These investigators (Porceddu *et al.*, 1999) considered introducing the human GAD65 (rather than GAD67) into transgenic plants because the former was supposed to play a greater role in the pathogenesis of human IDDM. These investigators thus inserted the cDNA for human GAD65 into a binary vector (pBin19) for *Agrobacterium*-mediated transformation. In this vector (pBin1935SGAD65), the cDNA for GAD65 was flanked by the 35S CaMV promoter and the *Nos* terminator. No signals were added to retain the GAD65 in the ER or in the organelles. Tobacco leaf discs and carrot hypocotyls were transformed and the regenerating shoots were selected (in kanamycin). Northern blot hybridization showed that seven out of nine tobacco plants had GAD65 transcripts. Further analyses were performed with transgenic tobacco plants having the highest transcript levels. Western blot hybridizations indicated that GAD65 was produced in the transgenic leaves. Interestingly, albeit no codes for organelle transit peptides were engineered in pBin1935SGAD65, immunogold labeling indicated that GAD65 was mainly located inside the chloroplasts and the mitochondria. The content of GAD65, evaluated by radioimmunoassays led to an estimate of up to 0.04% (of total soluble protein) of GAD65 in tobacco leaves and up to about

0.012% (of total soluble protein) of GAD65 in the edible roots of transgenic carrot plants.

There is no information on how much GAD65 would be required for repeated oral administration in man to suppress the autoimmunity of IDDM. But one can estimate that at the present level of GAD65 in transgenic carrot, a *lot* of carrot has to be consumed. If carrot roots contain about 0.1% of soluble protein (of wet weight), that means 1 kg of carrot will contain about 1 g of protein. Out of the latter, about 0.1 mg will be GAD65. If the daily requirement is 10 mg GAD65, then 100 kg of carrot will have to be consumed daily! So, a vast increase in production of GAD65 in carrots is required before this approach becomes feasible. The authors had several suggestions on how to increase the level of GAD65 in transgenic plants.

A somewhat different approach to recruit transgenic plants in order to furnish a remedy against IDDM was taken by Arakawa *et al.* (1998b). The rationale of these investigators was as follows. IDDM is an autoimmune disease in which there is a destruction of the insulin-producing pancreatic cells. Humoral and cellular mechanisms are involved in this destruction but the latter is probably primarily mediated by CD_4 and CD_8 T-cells. In the animal model for IDDM, i.e. the NOD mice, insulin is one of the autoimmunogens recognized by these T-cells, and a subset of activated CD_8 T-cells infiltrates the pancreatic islets. Repeated administration of large quantities of insulin may prevent the spontaneous autoimmune diabetes. In this connection, the nontoxic B subunit of cholera (CT-B) may be useful. CT-B may be used as an efficient carrier and cell internalizer when insulin is chemically conjugated to CT-B. The CT-B may facilitate the antigen (insulin) delivery because it will present the insulin to the gut-associated lympoid tissue (GALT) where the T-cells reside. This should happen because of the high affinity of CT-B for the cell surface receptor G_{m1} ganglioside, located on the cells of GALT. Hence, feeding a CT-B/insulin conjugate in sufficiently large quantities may alleviate IDDM. These investigators (Arakawa *et al.*, 1998b) thus constructed two expression vectors that were based on their previous (Arakawa *et al.*, 1997) expression vector (see Fig. 14). The new vectors contained either the coding sequence for the human preproinsulin instead of the code for CT-B in the previous vector; or the coding for a fusion protein: a leader, 5′ of the sequence for the CT-B, followed by the coding sequence for a flexible hinge tetrapeptide, Gly–Pro–Gly–Pro (to bridge the CT-B to the preproinsulin), and then the coding sequence for the human preproinsulin. The latter was terminated by the coding sequence for SEKDEL (the ER retention signal).

After *Agrobacterium*-mediated transformation of potato plants with either the vector that contained only the human insulin cDNA (INS) or with the vector that contained the coding for the fused protein (CT-B–INS), plants that contained the respective transgene were obtained, and the transgenes were transcribed and translated as expected. The authors estimated the level of insulin protein in transgenic potato tubers to be 0.05% of total soluble protein (it was not stated if this level was the same after transformation with vectors containing INS alone or CT-B–INS). The CT-B–INS fusion gene caused the production of oligomeric CT-B–INS and this oligomer had specific affinity for G_{m1} ganglioside as well as both CT-B and insulin antigenicities. Since only pentameric CT-B has affinity to G_{m1} ganglioside, it was estimated that the fusion protein existed as a pentamer. Tuber-derived and leaf-derived CT-B–INS fusion protein had the same biochemical and antigenic properties.

For analyzing the induction of oral tolerance, the authors used female NOD mice. The mice, at the age of five weeks, were divided into three groups:

1. mice fed transformed potato tubers;
2. mice fed tubers producing only insulin;
3. mice fed tubers producing the CT-B–INS fusion protein.

The mice were fed 3 g potato tubers once a week until the ninth week. The 3 g of tubers were estimated to contain 30 μg of insulin (Group 2) or 20 μg insulin (as part of the CT-B–INS of Group 3). The mice were sacrificed at ten weeks of age for antibody titer assay and histopathological analysis of pancreatic tissue; or the mice were monitored for six months for the development of diabetes. The level of IgG1 anti-insulin antibody was by far higher in the group of mice fed with the tubers that contained the CT-B–INS fusion protein than in mice fed with INS only (Group 2). The latter group's level of anti-insulin IgG1 was similar to that of the control (Group 1) mice. At the time of killing (ten weeks), none of the three groups showed diabetic symptoms but the histology indicated that control mice developed insulitis, and statistical evaluation indicated that there was a significantly lower insulitis in mice fed with CT-B–INS-containing tubers. Finally, when diabetes was evaluated, there was a clear reduction in the percentage of diabetic mice after feeding with tubers containing the CT-B–INS fusion protein (about 10% in CT-B–INS fed mice and 80% in control mice, at the age of 24 weeks; these "percentages" were based on ten mice per group). The authors concluded their report by suggesting that their study demonstrated that feeding microgram quantities of plant-produced insulin conjugated with CT-B can effectively suppress development of autoimmune diabetes in NOD mice. The authors did not speculate on the usefulness of CT-B–INS to suppress human diabetes.

4.3. Antigen Expression by Plant Viruses

The idea of recruiting plant viruses in order to express alien genes in plants emerged at about the same time when *Agrobacterium*-mediated plant transformation became feasible (see Sec. 2.2).

The rationale for the use of plant viruses for this purpose was based on several features of these viruses. In many cases, viral replication and their systemic spread in the host plant are very extensive and quick. In these cases, at about two to three weeks after infection, the level of the coat protein of the virus may reach about 40% of the total protein of the host plant. When the above-mentioned idea emerged, i.e. in the late seventies and early eighties, the manipulation of RNA sequences was technically problematic while manipulation of DNA sequences was already well established. Therefore, attention was directed to DNA viruses that infect plants. The caulimoviruses are plant viruses having a double-stranded DNA genome. Thus, the cauliflower mosaic virus (CaMV) became the virus of choice for delivering alien DNA into plants. Plants of the Cruciferae family can be infected by CaMV DNA through mechanical inoculation. The virus then spreads quickly reaching 10^6 viral genomes per cell, in the leaves of the host plant. The Cruciferae include many crop plants such as several species of the genus *Brassica*. After studying CaMV for several years, Hull (1978) of the John Innes Institute, stated: "Cauliflower mosaic virus is one of the most promising candidates as a vector DNA in genetic manipulation in plants." Well, as we shall see, not all "promising candidates" become champions. The torch of CaMV studies towards its use as DNA vector was passed from the John Innes Institute, Norwich, to the Friedrich Miescher Institute, Basel. Barbara and Thomas Hohn, and associates of the latter institute performed intensive studies on CaMV (see Lebeurier *et al.*, 1982). Due to studies in Norwich, Basel and elsewhere, the molecular biology of CaMV was elucidated. But, when it was attempted to use CaMV as a "Trojan horse" to express alien genes in host plants, problems emerged. One of these was that it was found that almost the entire genomic DNA of CaMV is essential for replication and spreading. Only a short viral sequence, encoding aphid transmission capability (*gene*II) could be replaced by another (alien) coding sequence. But the capacity to harbor an alien DNA sequence instead of *gene*II is limited: only 250 bp. The F. Miescher team did replace the *gene*II with a 240-bp sequence that encoded the bacterial dihydrofolate reductase (DHFR), and the DHFR enzyme activity was found in extracts of plants infected with the engineered CaMV (Brisson *et al.*, 1984). In addition to the above-mentioned constraint, there were other problems with viruses as vectors (van Vloten-Doting *et al.*, 1985, titled their publication: "Plant-virus-based vectors for gene transfer will be of limited use because of the high error frequency during viral RNA synthesis").

Thus, the T-DNA of *Agrobacterium* became the favorable (Trojan) horse for the stable integration and expression of alien genes in transgenic plants. But, as we shall detail below, for the transient expression of specific antigenic epitopes, several plant viruses were successfully employed. Moreover, in one respect, the CaMV system retained its "promising" position: its 35S RNA promoter became the most widely used promoter (CaMV 35S promoter) in genetic transformation of dicots. Several reviews covered this topic (e.g. Lomonossoff and Johnson, 1995; Beachy *et al.*, 1996; Porta and Lomonossoff, 1996; Porta *et al.* 1996; Johnson *et al.*, 1997; Spall *et al.*, 1997; Porta and Lomonossoff, 1998; Lomonossoff and Hamilton, 1999). Most of these reviews are from the John Innes Centre, where the concept was originally proposed (when it was still John Innes Institute), but then the main virus used for the delivery of antigenic epitopes was the cowpea mosaic virus (CPMV) rather than the CaMV.

The introduction to many of the publications that were dealing with the use of plant viruses for the production of antigenic epitopes contained a "promotion" component. It is claimed in these introductions that the conventional, stable genetic transformation is time consuming and is inflicted with several problems: some plant species are refractory to regeneration, the levels of expression of transgenes are often low and there are phenomena such as transgene silencing and co-suppression that adversely affect the expression of the transgenes. Well, in the commercial world, any merchant is allowed to criticize the merchandise of others and praise his own merchandise. But, we should remember that this has nothing to do with scientific endeavor.

4.3.1. *The early studies*

The first publication on the production of antigenic epitopes by a plant virus concerned an anti-polio vaccine (Haynes *et al.*, 1986). The plant virus was tobacco mosaic virus (TMV is a single-stranded RNA virus.) But, no plant infection was involved. These authors synthesized the coding DNA sequence for the coat protein of TMV (TMV-CP) and modified the 3′ end of this code by adding the coding sequence for a poliovirus type 3 antigenic epitope. The fused gene was termed TMV-CP–polio 3. The fused gene was expressed in *Escherichia coli.* The bacteria expressed this fused gene and the expressed transgenic protein assembled into TMV-like particles. This publication opened the possibility to use plant viruses, that have the capability of extensive multiplication in the appropriate plant host, for the production of antigenic epitopes.

The real breakthrough in the use of plant virus coat proteins as carriers of antigenic epitopes, that can serve as vaccines, followed progress in molecular biology methods. It became possible to synthesize cDNA for the viral coat protein and add the codes for specific epitopes to these cDNAs. Moreover, for several plant viruses, it was technically possible to use the DNA that encodes the modified viral genome (RNA) for infection of host plants and, by that, cause the multiplication and spread of the respective viruses in the host plants.

Donson *et al.* (1991) reported on an important step towards the use of single-stranded RNA plant viruses as carriers of alien genes that will be expressed in the host plant after viral infection. These investigators managed to engineer a cDNA that would result in the assembly of viable virus which will not only multiply at the infected site, but the virus will spread systemically in the host plant where it will also express the alien gene. It should be noted that previous attempts to use single stranded RNA plant virus for this purpose failed. Either the movement of the virus in the infected plant was impaired or the added alien genes were gradually deleted. The Donson–Dawson team (i.e. Donson *et al.*, 1991) had ample previous experience with tobamoviruses (to which TMV and the odontoglossum ringspot virus, ORSV, belong). The novelty of the new hybrid TMV vector was that the cDNA for the TMV was maintained up to (and including) the sequence encoding the TMV movement protein (30 kDa), and then another sequence was fused downstream. Three vectors were constructed. In the first vector, only the coding sequence of the ORSV coat protein with its own subgenomic promoter was added. In the second vector, the bacterial *dhfr* gene was added downstream of the code for the TMV 30 kDa movement protein and thereafter, the coat protein of ORSV and its promoter were inserted. In the third vector, the *npt*II gene (for kanamycin resistance) was inserted instead of the bacterial *dhfr* gene. The vectors were used to inoculate mechanically *Nicotiana benthamiana* plants. This inoculation resulted in the production of a hybrid virus that spread in the host plants and reached a high level. The *npt*II gene was partially deleted during serial passages but the *dhfr* gene was stable and was expressed. As shall be detailed in Chap. 5, this hybrid viral vector was used to produce α-trichosanthin in transfected plants (Kumagai *et al.*, 1993).

4.3.2. *The use of cowpea mosaic virus*

A study that was aimed specifically to express antigenic epitopes on the coat of a plant virus was reported by Usha *et al*. (1993). This was an international team of investigators then at Purdue University, Indiana, USA (the first author then moved from Indiana to India) and investigators from the John Innes Centre, Norwich, England (Lomonossoff and associates). They chose the cowpea mosaic virus (CPMV, a comovirus) as the carrier of the alien epitopes. The CPMV multiplies extremely well in the host plants (*well* for the virus and for those intending to utilize the virus as an epitope-presenting system, not so well for the host plant and the farmer who likes to have good crops of cowpeas...). The virus yield may reach 1–2 g per kg of infected plant tissue. The structure of CPMV is known. It is an icosahedral virus with a diameter of 28 nm. Its capsid contains 60 copies, each composed of a large (L) and a small (S) subunit. CPMV has some similarities to the mammalian picornaviruses (e.g. hepatitis A virus, foot-and-mouth disease virus and poliovirus). Each of the L proteins consists of two β-barrel domains that are positioned at the threefold axes of symmetry. The amino- and carboxy-terminal β-barrel domains of the L protein correspond to proteins VP2 and VP3, respectively, in a picornavirus. The S protein molecules that consist of a single β-barrel domain are packed around the fivefold axes and correspond to the VP1 of picornaviruses. The genome of CPMV consists of two separately encapsidated, positive RNA molecules of 5889 (RNA1) and 3481 (RNA2) nucleotides. The RNAs contain single open reading frames that are translated into two respective polyproteins that are subsequently processed (Fig. 15). Both RNA1 and RNA2 are required to initiate an infection in intact plants. But, RNA1 can replicate independently in infected protoplasts. The replication of RNA2, that encodes both of the viral coat proteins (L and S), is dependent on fractions furnished *in trans* by RNA1. The full-length cDNA of CPMV was constructed and can be transcribed *in vitro*. This provided the possibility of engineering this cDNA so that alterations can be introduced to the coding region of RNA2 and by that it was possible to manipulate the coat proteins. The modified cDNA could be used to infect plants where the virus could multiply and spread (see scheme in Fig. 16). Equipped with this knowledge and biotechnology, the authors (Usha *et al.*, 1993) aimed to introduce specific epitopes into the S protein. In this study, the investigators choose an epitope from the foot-and-mouth disease virus (FMDV). Thus, oligonucleotides that code for an epitope derived from VP1 (the "FMDV loop") of FMDV were cloned into a specific region of the RNA2 cDNA that codes for an exposed domain of the S protein. The alien coding sequence was introduced either as an addition or as a replacement. For infection, the *in vitro* modified RNA2 was mixed with RNA1 and the mixture was electroporated into cowpea protoplasts. After 72 hours, the replication of the modified RNA2 could be detected. Capsids were formed only if the whole S protein was produced: when part of it was *replaced* by the FMDV, no capsids were formed. The CPMV that had an insertion of the FMDV epitope could multiply after cowpea plants were infected with the respective sDNAs, but there was no systemic spread of the modified CPMV. Infected leaves were analyzed by immuno-absorbent electron microscopy revealing mostly empty capsids and also a few capsids with RNA. Western blot hybridization of protein from leaves of plants infected with the modified CPMV showed that this protein reacted with antiserum against the inserted FMDV. In short, while there was some production of the CPMV that carried the FMDV epitopes, the infection did not lead to massive production of the modified virus. The modified viral genome was apparently not stable.

In a further study by this team, (Porta *et al.*, 1994), it was revealed that there is a tendency for the modified viral genome to lose in one step the entire code for the FMDV epitope. The investigators then redesigned the chimeral genome to render it genetically stable. The change was to insert the alien epitope at a different location of the S coat protein of the CPMV. Thus, cDNAs that had either the code for an epitope of the human rhinovirus-14 (HRV-14) or human immunodeficiency virus type 1 (HIV-1) were constructed. These codes were now inserted into the new site of the code for the S of CPMV. These chimeral cDNAs were transcribed *in vitro* and each of the two transcripts was mixed with the transcript for RNA1 of CPMV and used to mechanically inoculate cowpea plants. The chimeric virus particles that resulted from this inoculations had the expected antigenic properties. When plants were inoculated with a mixture of RNA1 and the chimeral transcript coding for the HRV-14 epitopes in the modified CPMV, the resulting virus was immunogenic in rabbits. It should be noted that quite high levels of the chimeric CPMV could be obtained in infected cowpea plants: about 1 mg of chimeral CPMV per g of leaves.

A plant-virus-derived vaccine was developed by a team of 15 (!) investigators (from six institutions), against three other animal viruses: canine parvovirus (CPV), mink enteritis virus (MEV) and feline panleukopenia virus (FPLV). These investigators (Dalsgaard *et al.*, 1997) identified a linear neutralizing epitope at the amino terminus of the CPV VP2 capsid protein. This amino acid sequence is shared by CPV, MEV and FPLV. The code for this amino acid sequence was cloned into the cDNA that codes for the S coat protein of CPMV

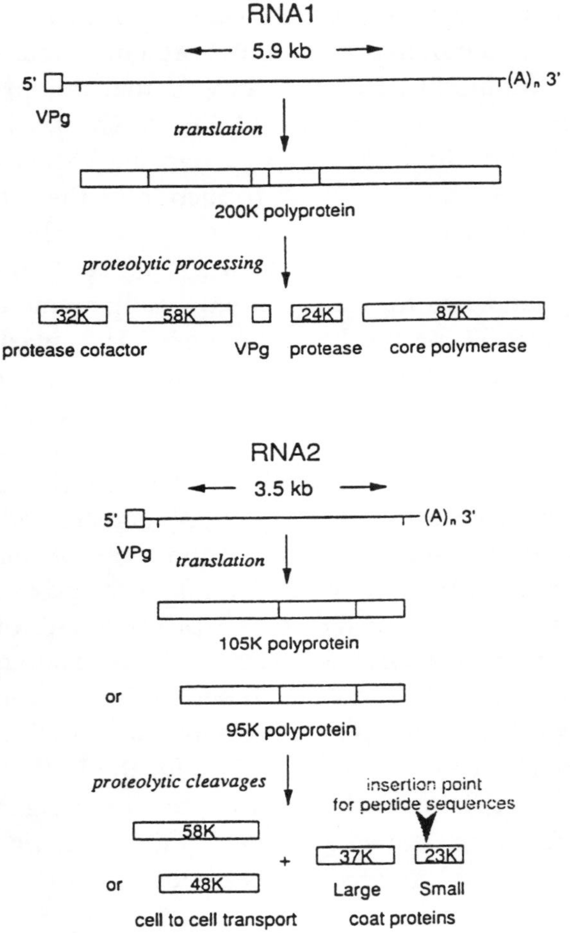

Fig. 15. Diagram of the CPMV genome and its expression products. Each of the two genomic RNAs contains a single long ORF and is translated into a large polyprotein. Maturation of this precursor protein occurs by cleavage with virus-encoded proteases. In the case of RNA-2, both a 105 KDa and a 95 KDa protein are produced as a result of secondary initiation downstream of the first inphase AUG codon. RNA-1 encodes the replication-associated functions and RNA-2 codes for the cell-to-cell movement protein and the two coat proteins. Insertion of foreign peptide sequences have been made in the small coat protein as indicated by the grey arrow. (From Porta and Lomonossoff, 1996.)

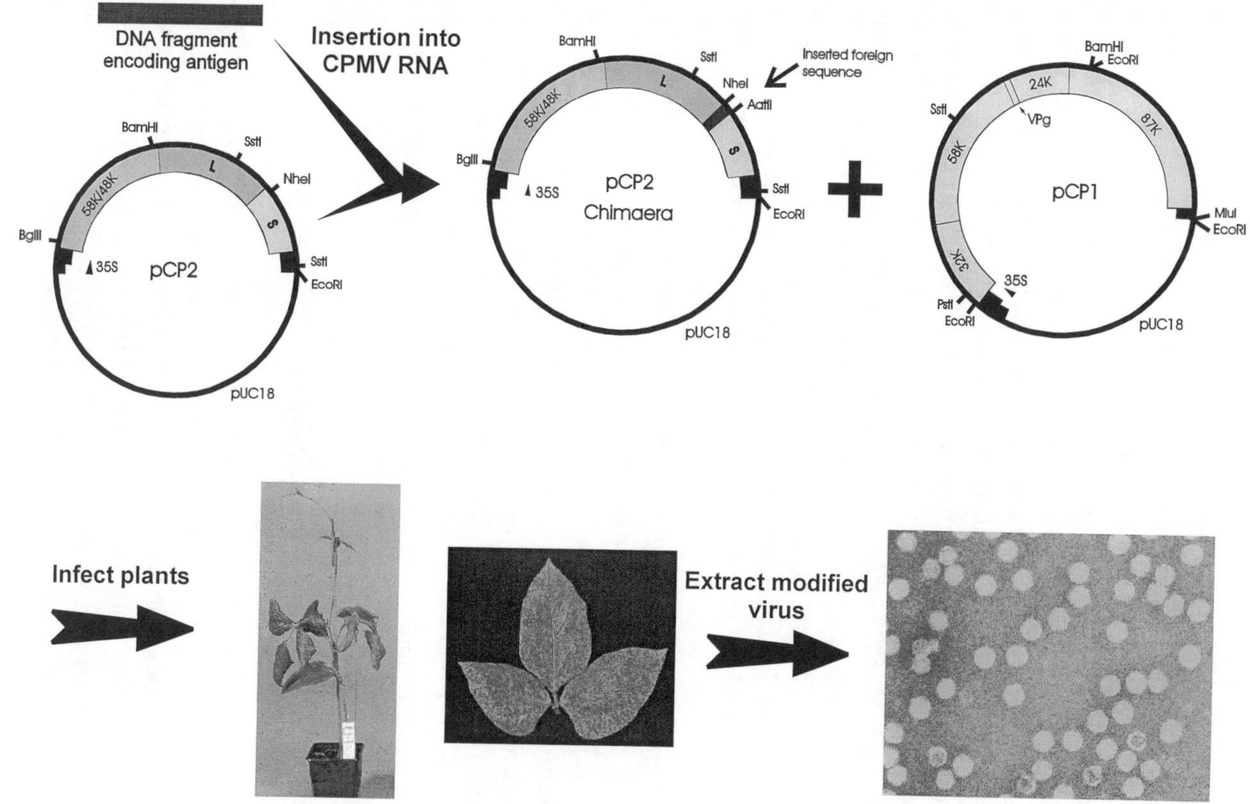

Fig. 16. Scheme of the method to construct and propagate CPMV chimeras. (From Lomonossoff and Hamilton, 1999, where further details are provided.)

between nucleotides 2725 and 2726 of the S coat protein cDNA sequence. The respective plasmid (pCP2-Parvol) was inoculated together with the unmodified plasmid (pCP1) into *Vigna unguiculata* (cowpea) plants. All five inoculated plants showed the typical CPMV symptoms seven days after inoculation. After five more days, the symptoms appeared in the leaves above the inoculated leaves. Then, there was the typical spread of the symptoms. Five inoculated plants yielded about 50 mg of virus (from about 50 g leaf material). The modified virus was stable for at least two passages. Electron microscopy revealed CVP-Parvol particles that were similar to wildtype CPMV. SDS–PAGE and Western blotting confirmed the existence of two S coat proteins: the wildtype and the modified CP. Doses of either 100 μg or 1 mg were mixed with adjuvant (Quil A and aluminium hydroxide) and were injected into groups of six mink animals (6–8 months old) that were infected with MEV. The vaccination was applied once subcutaneously. The lower dose prevented MEV in five out of six animals and the higher dose prevented MEV in all six animals. Also, the higher dose prevented MEV shedding in the feces. Control minks, infected with MEV but not immunized with CVP-Parvol, suffered severe diarrhea. While, in this publication, the CVP-Parvol was not used to protect dogs and cats against CPV and FPLV, the authors expect that such a protection is likely (and will be tested in the future) because, in all three viruses, the same amino acid sequence was used for the vaccination of the minks against MEV.

4.3.3. *Focus on human immunodeficiency virus (HIV)*

During the recent decade, there was hardly any other human malady that attracted the public and medical profession more than the human immunodeficiency virus type 1 (HIV-1), the forerunner of AIDS. One of the approaches to treat HIV patients is to cause an effect on the immune system of these patients in order to at least suppress the HIV virus for a certain period. Some hints that this could be achieved was the observation that antibodies directed against certain epitopes (e.g. gp120 and gp41) on the envelope protein of HIV-1 could neutralize infectivity *in vitro.* But obtaining *in vivo* such neutralizing antibodies was difficult. AIDS is an important and still incurable disease. Therefore, it focused the attention of investigators who had in mind of furnishing specific antigens that will elicit the required antibodies by epitopes on the coat protein of plant viruses. A team which included experts on HIV, experts on the expression of antigens on plant virus coat proteins and an investigator from a commercial company joined forces in this endeavor (McLain *et al.*, 1995). These investigators chose the capsid of CPMV to express the antigenic epitopes that will elicit the respective antibodies. They also made use of their previous experience with CPMV in order to insert the antigenic epitope at the right site. They thus used a similar technique (see above) to construct a cDNA for the CPMV RNA-2 which codes for the S coat protein (as well as for the L protein). The oligonucleotide sequence that codes for amino acids 731 to 752 of the gp41 envelope protein of HIV type 1 IIIB was cloned into the correct place of this cDNA. The transcript for this cDNA was mixed with the RNA-1 transcript and the mixture was used to transfect young cowpea (*Vigna unguiculata)* plants. The leaves of the infected plants were homogenized in 0.1 M sodium phosphate, the homogenate was cleared by centrifugation and the supernatant was used for a second round of inoculating cowpea plants. This provided a large quantity of modified CPMV (that furnished modified virus for the next three years!). Purified virions extracted from the inoculated leaves maintained the structure of CPMV and analyses indicated that they contained the modified S protein. The injection of these virions (with alum adjuvant) subcutaneously into mice elicited neutralizing antibodies against HIV in the mice. The antisera were apparently specific against the gp41 oligopeptide. The neutralizing activity declined a few weeks after the second injection of modified CPMV but could be raised to some extent by the third injection, and then declined again. The investigators noted that some neutralizing activity was also elicited by the unmodified CPMV virions. This latter neutralizing activity could be removed by adsorption with purified CPMV, indicating the (low) specificity of the anti-HIV-1 antibody response.

The stimulation of neutralizing antibodies to HIV-1 was further studied by this team (McLain *et al.*, 1996). In this study, it was investigated whether or not different murine lines differ in this neutralizing antibody formation and if the immunization with the same modified CPMV (now termed CMV–HIV/I) can have a longer lasting effect. Three murine lines were injected with 100, 10 or 1 µg chimeric virus. The higher doses (100 or 10 µg) caused close to 100% neutralization of HIV-1 (IIIB). The 1µg dose gave lower neutralizing values. But, there was an unexplained phenomenon; the longevity of the neutralizing effect after a 1 µg injection was much greater than after a 10 or 100 µg injection. Actually, after a 100 µg injection, the effect disappeared rapidly.

A third study (Durrani *et al.*, 1998), with the same modified virus (CPMV–HIV/I), tested intranasal and oral immunizations. The rationale was that for HIV, it is an advantage to have mucosal antibodies because the mucosa is at the infection site (e.g. the vagina). The CPMV–HIV/I was mixed with CT (cholera toxin) to enhance the immune response. Volumes of 10 µl containing 10 µg of modified virus were applied to the nostrils of mice. The oral doses were provided by gastric lavage. The levels provided per mouse for oral application were 100 to 1000 µg modified virus. Immunizing mice intranasally did cause production of HIV-1 gp41-specific fecal IgA (the authentic CPMV did not cause such a production). Some mice even responded after a single nasal application (10 or 100 µg chimeric virus) while all mice responded after a nasal boost. A third nasal application boosted the response only marginally. All mice immunized intranasally also produced serum antibody against the gp41 peptide (mostly IgG2a, less of IgG1). The effect of oral application was much lower than the effect of intranasal application even though the latter doses were about 100-fold lower than the doses of the oral applications. Only a minority of the mice that received the oral application developed serum immunoglobulin against HIV-1 gp41. While the way to medical use of the CPMV–HIV/1 virus is still very long, this study showed that intranasal immunization with the CT as adjuvant can induce specific IgA in the mucosa even when applied in very low doses (e.g. two doses of 10 µg modified plant virus). Although the IgA response was lower than that elicited by subcutaneous immunization, the intranasal application, possibly in combination with injection, may be considered in the future.

The University of Warwick and Axis Genetics teams (McInerney *et al.*, 1999) analyzed several adjuvants in this system. Each of the five adjuvants was administered subcutaneously, together with 10 µg wildtype CPMV or 10 µg of the chimeric CPMV (that displays the HIV-1 gp41 residues 731–752). The adjuvant Quil A elicited the highest and most consistent response to the modified CPMV, with respect to ELISA titers found with immobilized peptide and HIV-1 neutralizing antibody. The commercial adjuvant Quil A (Superfos, Denmark) was also the only adjuvant to stimulate an *in vitro* proliferative T-cell response.

A recent study, with an epitope (Nlm-1A) from the human rhinovirus-14 (Taylor *et al.*, 1999) indicated that alterations in the "graft" of the epitope to CPMV can improve the antigenicity of the modified CPMV coat protein.

A different virus, tomato bushy stunt virus (TBSV), was used by a team from Uppsala, Sweden and the University of Nebraska, Lincoln, USA, to develop a vaccine against HIV-1 (Joelson *et al.*, 1997). These investigators focused on a 13-amino-acid peptide derived from the V3 loop of HIV-1 glycoprotein 120 (gp 120). They constructed a modified TBSV in which the 13 amino acids were attached to the C-terminus of the coat protein. If all the subunits of the icosahedral virus would express this epitope, it would have been repeated 180 times in each viral particle. The infective virus can be transcribed *in vitro* from a full-length cDNA. The TBSV genome is a single positive-sense RNA virus having the length of 4,776 nucleotides, and the genome has been sequenced. Thus, for insertion of the 13 amino acids into the CP subunits, the coding sequence for these was cloned at the 3′ end of the code for the CP (Fig. 17). The RNA transcript of pTBSV-STV3 was used to inoculate *Chenopodium amaranticolor* and *Nicotiana benthamiana* leaves resulting in the typical symptoms as wildtype TBSV. Leaves of inoculated *N. benthamiana* were used to purify the modified virus and reinfect further plants. The viral titers of wildtype-inoculated and STV3-inoculated leaves were similar, about 1 mg virus per g of leaves. The viral particles were also similar when observed by electron microscopy (Fig. 18). The modified virus was stable after several passages in plants (after six passages, there was one exchange: isoleucine to methionine). The HIV epitopes of the TBSV-STV3 were detected by V3-specific monoclonal antibody and by the sera of HIV-1-positive human patients. The chimeric virus

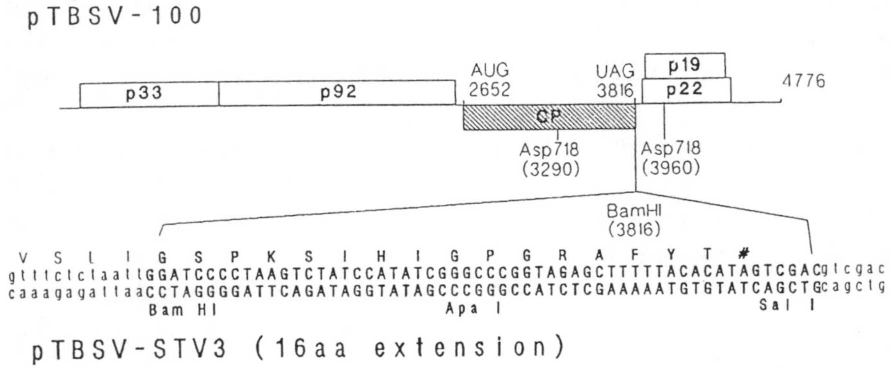

Fig. 17. Diagram showing the genomic organization of tomato bushy stunt virus (TBSV) and the derived mutant pTBSV-STV3. The open boxes represent the genes identified in Hearne *et al.*, *Virology* **177**, 141 (1990). The capsid protein gene is designated CP. The positions of the two Asp-718 restriction sites used in this study and the newly introduced sites (*Bam*HI, *Apa*I and *Sal*I; included to facilitate future cloning experiments) at the end of the CP gene are identified. The nucleotide sequences derived from HIV-1, and those corresponding to the three linker amino acids (GSP), respectively, cloned at the end of the coat protein gene are in bold, with the adjacent TBSV nucleotide sequences in lower case. The amino acid sequence of the insert is in bold above the nucleotide sequence. (From Joelson *et al.*, 1997.)

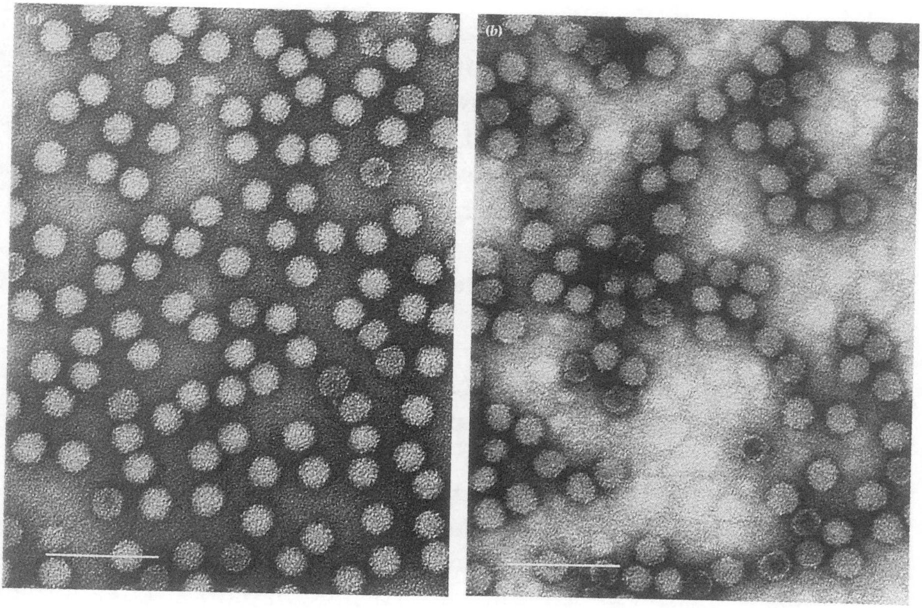

(a) (b)

Fig. 18. Electron microscopy of (a) wildtype TBSV and (b) TBSV-STV3. Scale bar: 100 nm. (From Joelson *et al.*, 1997.)

also induced a specific immune response when injected into mice. These findings suggested future use of TBSV-STV3 for diagnostic purposes and possibly for the development of an HIV vaccine.

The alfalfa mosaic virus (AIMV) coat protein was also utilized to express antigenic epitopes from HIV (as well as from rabies virus). Yusibov *et al.* (1997) chose AIMV coat protein because it can accommodate larger alien peptides. The AIMV comes in different shapes — spherical, ellipsoid and bacilliform, according to the length of the encapsidated RNA. The N terminus of the AIMV CP is located on the surface of the virus particle and thus apparently does not interfere with the correct folding. Yosibov *et al.* (1997, 1999) engineered the respective cDNAs (with codes for epitopes of either HIV-1 or rabies virus) by cloning the modified AIMV CP downstream of TMV cDNAs.

The resulting cDNAs were transcribed *in vitro* and used to inoculate tobacco plants. The two types of viruses: TMV–AIMV-CP with rabies virus epitopes and the TMV–AIMV-CP with HIV-1 epitopes were isolated from the respective inoculated tobacco plants. Immunization of mice with these antigenic viruses elicited specific virus-neutralizing antibodies against the respective rabies virus and HIV-1.

4.3.4. *Production of antigenic epitopes in tobacco mosaic virus (TMV)*

We noted above that Donson *et al.* (1991) devised a vector that induced, in transfected tobacco plants, the production of modified TMV. This vector was based on a chimeral cDNA that included a TMV–odontoglossum ringspot virus (ORSV). The purpose was to avoid RNA recombination in the modified TMV–ORSV. A team composed of Hamamoto, Sugiyama and collaborators (of the Kanebo Ltd., Kaganawa and Teikyo University, Tochigi, Japan) used another construction to obtain a chimeric TMV (Hamamoto *et al.*, 1993). They used the information that a six-base sequence that follows the stop codon for the TMV 130 kDa protein gene permits readthrough of the stop codon. Hamamoto *et al.* (1993) thus inserted these six bases behind the stop codon for the TMV CP gene and immediately 3′ of these, the transgene was inserted. In a further study (Sugiyama *et al.*, 1995), this team used this approach to obtain TMV with modified CP that will express alien epitopes.

The alien epitopes were from influenza virus hemagglutinin (HA) or from HIV-1. Because the alien epitopes were placed at the C terminus of the CP, they should protrude from the rod-shaped (modified) TMV. Three such vectors were constructed: two to express HA epitopes (pTLHA1 and pTLHA2) and one to express the HIV-1 epitope (pTLHIV). These vectors coded for 18 or 8 amino acids of HA (pTLHA1 and pTLHA1, respectively) or for 13 amino acids of HIV-1 (pTLHIV). Because of the leakiness of the stop codon, each vector induced the production of the CP with the alien epitope as well as the authentic TMV-CP. Each TMV virion contains 2,130 spirally arranged subunits; therefore, even if only a small percentage of these subunits carry an exposed alien epitope, each virion should contain many such epitopes. The investigators synthesized the respective three RNA vectors and encapsidated them *in vitro* with CP. The three types of modified TMVs (as well as control TMV) were used to inoculate leaves of young tobacco plants. This resulted in the respective mosaic symptoms (after 2–3 weeks) in the upper (not inoculated) leaves of the tobacco plants. Analyses indicated that these upper leaves contained the modified CP. This was revealed by Western blot hybridization with anti-TMV antiserum. While the TMV extracted with control (TMV) showed only the TMV-CP band, the leaves of plants infected with pTLHA1, pTLHA2 or pTLHIV had an addition band corresponding to the respective modified CPs (with the additional 18, 8 or 13 amino acid peptides from HA or HIV). The virus particles of authentic TMV are resistant to trypsin treatment but when the modified TMVs were treated with trypsin, the epitopes were removed. Thus, their size, as revealed by electrophoresis, was reduced by this treatment to the size of authentic CP of TMV. This removal by trypsin was possible because the added epitope contained either one or both arginine and lysine, close to their N-terminal side. These results thus indicated three important issues:

— The added epitopes occurred in some of the CP subunits of the TMV;
— The added epitopes did not prevent the normal folding of the CP;
— The added epitopes were on exposed locations of the CP.

We did not see, till presently, reports on further work on this subject by these investigators.

Fitcher *et al.* (1995) recruited the ability of modified CP of TMV to express alien epitopes. The epitopes they handled were oligopeptides that may elicit antibody which interfere with the murine *zona pellucida*, inhibiting the sperm binding to the egg. Such an inhibition could lead to antibody-mediated contraception. These investigators used the common strain of TMV. By appropriate engineering of the respective cDNAs, the coding sequence (39 nucleotides) for an epitope was inserted into the coding region for the CP of the TMV. The insertion was performed so that the code for 13 amino acids of the murine *zona pellucida* ZP3 protein ($ZP3_{331-343}$) was added between the codes for amino acids 154 and 155 of the CP. They thus obtained a full-length cDNA for the modified TMV. This cDNA was transcribed *in vitro* and the resulting RNA was used for inoculating tobacco leaves. The transcription product was diluted (tenfold) in 20 mM sodium phosphate (pH 7.2) and 100 ml of this dilution was applied to each expanding leaf by gentle rubbing in the presence of carborundum powder. To produce large quantities of modified TMV, crude homogenates of infected leaves were used to infect many new (uninfected) leaves. The modified TMV was then extracted form the upper leaves of infected plants. The infection with the hybrid RNA caused the typical symptoms of TMV infection. The symptoms appeared two weeks after infection. The modified CP was monitored by SDS–PAGE in which the modified CP migrated slightly slower than wildtype CP. The modified TMV also spread systemically (as in wildtype) from the infected leaves to the upper leaves. Western blot hybridization of extracts from leaves of plants that were infected with the modified RNA (as well as control extracts), with anti-$ZP3_{336-342}$ monoclonal antibody revealed that the leaves of plants infected with the modified TMV contained the modified CP. No hybridization band was revealed in control extracts. Mice (of two lines) were injected (100 or 200 μl per injection) with 50 μg of either wildtype or modified TMV-CP. The injections were applied into the peritoneal cavity and subcutaneously into the neck of BALB/c mice. The mice reacted rather uniformly and 21 days after the first injection, anti-$ZP3_{331-343}$ antibodies were elicited in these mice. Cryosectioning of the *zona pellucida* of mice injected with the modified CP (but not of control mice) indicated that antibodies, mostly IgG, were recruited to the *zona pellucida*. Noteworthy was the reduced size of the ovaries in mice injected with the modified TMV-CP.

Finally, regarding the fertility of female mice treated with the modified CP, the results were disappointing. Modified-CP-treated female mice and control mice were caged for ten days with male mice. There was no reduction in pregnancy in the treated mice as compared to the control female mice.

One may question whether or not the whole attempt was worthwhile. Even if the treated mice could have shown a reduced fertility (or no fertility at all) and no adverse effect on the mouse ovaries would have occurred, we do not see a good reason to go through such a complicated process of autoimmunity in order to achieve contraception. After all, there are simpler means of contraception in human females that are almost devoid of risks, are rather effective and are reversible. In short, the poor female mice; we remember the "three blind mice" from the "*Mother Goose*" verses.

Further studies by R. Beachy and associates on the expression of immunogenic epitopes on the CP of TMV revealed interesting information. One of these studies (Bendahmane *et al.*, 1999) solved an enigma. It was previously found that the insertion of certain peptides in the TMV-CP, at or close to the C end of the CP may cause problems. The modified TMV was either not moving away from the site of infection or the viral assembly was impaired. By genetic engineering, these authors came up with CPs that contained a number of peptides, inserted between Ser 154 and Gly 155, or following the last amino acid (Thr 158) of the CP. In the respective hybrid CPs, the isoelectric points (pIs) of the peptides ranged from 3.84 to 12 and the *pI: charge* values also varied. It was found that in cases where the pI and the pI: charge of the hybrid CP were beyond a certain limit, there was no viral movement and/or no viral particles were formed. There was an important conclusion: the pI of the peptide–CP fusion protein influences the interaction of TMV with the host. The TMV-CP tolerates negatively charged peptides but is negatively affected by positively charged epitopes. This situation could be corrected by the exchange of some amino acids of the epitope. But, of course, these changes should be compatible with the immunogenicity of the epitope.

In a subsequent study, this team (Koo *et al.*, 1999) dealt with protective immunity against murine hepatitis virus (MHV) by chimeric TMV that has immunogenic peptides exposed as epitopes on its CP. MHV, a member of the Coronaviridae family, is responsible for a variety of acute and chronic diseases in its natural host. It is the nightmare of keepers of experimental animals. When MHV epidemic breaks out in an installation for breeding and keeping of such animals (e.g. mice), the installation has to be evacuated and disinfected thoroughly. One of the dominant structural proteins of MHV is the spike glycoprotein — *S protein*. The S protein is a critical determinant of viral pathogenicity and has major immunodominant neutralization domains. The S protein is post-translationally processed to yield the S_1 and S_2 proteins. Protective neutralizing antibodies were induced in mice by immunization with a synthetic peptide derived from the S_2 protein. Specifically, a monoclonal antibody 5B19.2 was shown to protect against lethal challenge by passive immunization. This 5B19.2 was used to map its recognition site and a sequence of a 10-amino-acid segment was revealed on the S_2 glycoprotein of an MHV strain (A59). This sequence is L^{900}LGCIGSTCA^{909} (the 5B19 peptide). Equipped with this information, Koo *et al.* (1999) questioned the possibility of obtaining a viable and infective modified TMV in which the 5B19 will appear as an epitope on the CP, and if such a modified CP could serve for active immunization of mice against MHV.

These investigators actually constructed two hybrid TMV-CP cDNAs. In both cases, the codes for the epitopes were inserted between the codes for amino acids Ser 154 and Gly 155. In one case, the peptide was the regular 5B19 (L^{900}LGCIGSTCA^{909}) and in the other case, it was P^{899}LLGCIGSTAEDGN^{913} (and termed 5B19L). By appropriate engineering, two respective cDNAs for the modified TMV-CP were obtained. These were transcribed *in vitro* and used to infect leaves of tobacco plants (Xanthi nn and Xanthi NN). The infection of Xanthi nn with TMV wildtype, TMV-5B19 and TMV-5B19L caused the typical TMV symptoms and systemic spread of the virus was observed. Infected leaves could be used to mass-infect additional tobacco plants but there was a major difference between TMV wildtype and TMV 5B19L versus TMV 5B19. The latter infection caused 90% of the TMV 5B19 to be included in the insoluble fraction. Only ultrasonication could solubilize the TMV 5B19 from the leaf extracts. Due to lack of availability of TMV 5B19, further studies were focused on the TMV 5B19L. The TMV wildtype and the TMV 5B19L were revealed in the leaves by appropriate analyses and used to immunize mice by schedules of either intranasal immunizations or subcutaneous immunizations. The intranasal immunizations caused the development of serum IgG and IgA, specific for 5B19 (and for the TMV-CP). Subcutaneous immunization elicited high titers of the respective serum IgG but not of IgA.

Mice that were first immunized with TMV wildtype or with TMV 5B19L (intranasally or subcutaneously) were then given a challenge of a lethal dose ($10 \times LD_{50}$) of MHV. All mice immunized with the TMV wildtype died after ten days. Mice that were immunized with TMV 5B19L were protected. With reference to the previous publication (Bendahmane *et al.*, 1999), the pI of 5B19L is 3.84 while the pI of 5B19 is 5.94. There is a take-home lesson for all those who intend to utilize antigenic epitopes on the CP of TMV: the pI and the pI: charge values should be taken into consideration. It is plausible that additional, not yet revealed, structural and chemical features of the hybrid CP affect the immunogenic efficiency.

4.3.5. *Plant virus epitopes as antigens for bacterial pathogens*

The bacterium *Pseudomonas aeruginosa* is an important pathogen. It is an opportunistic pathogen that infects newly born human infants, immunosuppressed patients and children with cystic fibrosis, and may cause chronic pulmonary infections. In rat models, efficient active immunization against *P. aeruginosa* was performed by using the outer membrane protein F of this bacterium as antigen. The use of such an antigen is not suitable for human vaccination. Thus, a team of investigators (from Axis Genetics, Cambridge, UK; Louisiana State

University, LA, USA and the Scripps Research Institute, La Jolla, CA, USA) studied the use of CPMV-CP as carriers of epitopes from protein F for the production of effective antigens (Brennan *et al.*, 1999a). Their approach was based on previous knowledge that a peptide representing a surface-exposed epitope termed *linear B-cell* epitope (not to be confused with the Minoan *Linear B* script, deciphered by the English school boy Michael Ventris) can be used to protect mice from *P. aeruginosa* infection. The synthetic peptide was tested as a substitute for the natural bacterial epitope since the latter is not suitable for human vaccination. Two plasmids were engineered: pCP2-PAE-4 and pCP3-PAE-5. PAE-4 contained the code for the S protein of the CPMV-CP into which the coding sequence for two *P. aeruginosa* epitopes (#10 and #18, with some modifications, total of 37 amino acids), were inserted. The other plasmid, PAE-5, was similarly engineered but contained the code for the modified L subunit of the CPMV-CP. These plasmids were used, as in previous studies of this kind, to infect cowpea plants. Also, similar to previous studies with CPMV epitopes, the leaves that were infected in a first round were used for further infection of cowpea plants to furnish large quantities of modified CPMV. The yields for CPMV-PAE4 and CPMV-PAE5 were about 1 mg per g of leaf tissue. Indeed, after infection with either of the modified CPMVs, the epitopes were found to be expressed. Subcutaneous injections (several) with adjuvants to mice indicated that CPMV-PAE4 induced mainly antibodies against peptide #10. The anti-peptide antibody elicited by CPMV-PAE5 was of the IgG_{2a} isotype. These IgG_{2a} antibodies were found to bind complement and to augment phagocytosis of *P. aeroginosa* by human neutrophils.

Brennan and associates also handled vaccination by plant virus (CPMV) epitopes against *Staphylococcus aureus* in two publications (Brennan *et al.*, 1999b; 1999c). *S. aureus* is another very important human pathogen. It inhabits the skin and certain mucus epithelia of man and other mammals. *S. aureus* is a major cause of nosocomial bacteremia, resulting in serious infections such as osteomyelitis, invasive endocarditis and septicemia; it may also cause peritonitis in patients undergoing continuous ambulatory peritoneal dialysis. In short, one likes very much to get rid of *S. aureus*.

Brennan *et al.* (1999b) reported on nasal application of modified CPMV that contains on its CP antigens against the D2 peptide of the fibronectin-binding protein B (FnBP-B) of *S. aureus*. These authors engineered a plasmid, in a similar manner to their previous construct, that will result in a cDNA for a CPMV with a modified CP. The code for the 30 amino acids of the D2 domain of FnBP-B was inserted into the code for the S CP of the CPMV (between the codons for amino acids 22 and 23). The resulting plasmid pCP2-MAST1 was linearized and mixed with linearized pCP1 (encoding the other RNA strand of CPMV required for viral multiplication). The mixture was used to inoculate young cowpea plants and the modified CPMV of the first inoculation was used to inoculate many leaves of young cowpea plants from which the stock of modified virus was obtained. Each of the virions was estimated to carry 60 copies of the D2 peptide and the yield of modified CPMV was about 1 μg per g leaf tissue. Thus, it was calculated that 1 μg of CPMV-MAST1 contained 40 ng of D2 peptides. For intranasal application, control CPMV or CPMV-MAST1 (with or without adjuvant) was applied to lightly anesthetised mice in doses of 100 μg (in four applications at seven days intervals). For subcutaneous immunization, mice were given 10 μg of wildtype or CPMV-MAST1 (with adjuvant) at days 0 and 14. For oral immunization, mice were given 100 μg of virus with adjuvant (in 100 μl). The nasal immunization was the most efficient one and did not require adjuvant. After four applications of CPMV-MAST2 (at day 50), high levels of FnBP D2-specific IgG were found in the sera of the mice and D2-specific IgA in the bronchial, intestinal and vaginal lavage fluids. After nasal immunization with wildtype CPMB, no such antibody was detected. Adjuvant (ISCOM matrix) did not increase the elicitation of antibody. The IgGs in the sera of mice immunized intranasally with CPMV-MAST1 were predominantly of the IgG2a and IgG2b subclasses. Purified IgG from the sera of immunized mice was found to inhibit the binding of human fibronectin to immobilized FnBP but this inhibition was less than the inhibition by IgG of mice immunized subcutaneously. The oral immunization was much less effective than the intranasal immunization. The authors concluded that the modified CPMV is an efficient carrier of foreign peptides to the mucosal immune system. Intranasal vaccination of this carrier causes efficient elicitation of mucosal and serum antibodies and could provide a protective response against pathogens.

In another publication (Brennan *et al.*, 1999c), these investigators made a comparison between two plant viruses as carriers of the D2 peptide derived from the *S. aureus* FnBP. They compared the capability of CPMV, used in their previous study, to express an amino acid sequence from the *S. aureus* fibronectin-binding protein (FnBP), to the capability of another plant virus, the potato virus X (PVX), to express this sequence. The latter virus can express approximately 1500 copies of a foreign peptide per virus particle as compared to 60 copies of such a peptide by CPMV. As the cDNA for the synthesis of CPMV-MAST1 was already available (see above), the authors constructed an analogical cDNA based on the PVX. The cDNA was constructed in a way where all the 38 amino acids of the D2 from the *S. aureus* FnBP will be expressed on the PVX CP subunits. The resulting plasmid was linearized and used to infect *Nicotiana benthamiana* leaves. This resulted in the production and spread of the modified PVX that was termed PVX-MAST8. The yield of modified PVX was about 0.2 mg per g leaf tissue (as compared to about 1.2 mg per g for CPMV-MAST1) and was calculated to contain 15 ng alien peptide per mg of PVX-MAST8. Mice were given 1, 10 or 100 μg of virus (PVX-MAST8, PVX, CPMV-MAST1 or CPMV) subcutaneously (with adjuvant) and then followed by one or two booster injections. In one experiment, rats were immunized subcutaneously with 1 mg of CPMV-MAST1 (with adjuvant) followed by two injections (days 14 and 28). An ELISA assay indicated the presence of the epitopes in CPMV-MAST1 and PVX-MAST8. The immunized mice that were injected with either PVX-MAST8 or CPMV-MAST1 produced high titers of FnBP-specific antibody. The mice IgGs were mainly of the IgG_{2a} and IgG_{2b} isotypes that strongly bind complement component C1q, suggesting a TH1 bias in the peptide-specific response. The sera of mice that were immunized with CPMV-MAST1 or with PVX-MAST8 were found to inhibit the binding of fibronectin to immobilized recombinant FnBP. These results indicated that CPMV and PVX can be used as carriers of immunogenic epitopes. The authors predicted that both plant viruses are useful to develop vaccines to protect against *S. aureus* infection.

A further publication on the use of antigenic epitopes on the CP of a plant virus came from another team (Staczek *et al.*, 2000). These investigators also handled *P. aeruginosa*. Here, we note that *transitions* are quite frequent in the realm of plant biotechnology. The Axis Genetics members of the Brennan *et al.* (1999a) team were meanwhile dissolved as Axis Genetics near Cambridge, UK, does not exist anymore. Furthermore, Roger Beachy (of the teams of Bendahmane *et al.*, 1999 and Koo *et al.*, 1999, mentioned above) moved eastwards from La Jolla, CA to the Midwest, to become President of the Donald Danforth Plant Science Center in St. Louis, MO. Previously, the Brennan *et al.* team included H.E. Gilleland Jr. of the Louisiana State University, but the same Gilleland Jr. was also coauthor in the publication by Staczek *et al.* (2000). Complicated? Perhaps it is, but it is of no scientific relevance. The only investigators of *P. aeruginosa* who did not move geographically (i.e. the Gillerlands), made the methodological *transition* from the use of CPMV to the use of TMV. Staczek *et al.* (2000) stressed in their abstract: "This is the first example of TMV being used to construct a chimera containing a bacterial epitope." Well, this is surely true, especially if we recall that the original *chimera* (of the Greek mythology) was composed "only" of a lion's front, a serpent's back and a goat's in-between (no bacteria were involved). Now, back to our topic. The outer membrane (OM) of *P. aeruginosa* contains protein F that has demonstrated vaccine efficiency against this bacterium in animal models. Vaccination of rats or mice with synthetic peptides representing surface-exposed epitopes of OM protein F identified two discontinuous epitopes (represented by synthetic peptides 9–14 mer and 10–14 mer) from near the carboxyl terminus of OM protein F. These elicited antibodies reactive with the whole cells of heterologous immunotype strains of *P. aeruginosa* and also elicited a sufficient immune response so as to significantly decrease morbidity and mortality of infected mice and rats. Staczek *et al.* (2000) thus attempted to use the 9–14 mer peptide as epitopes on the CP of TMV in order to develop a vaccine against *P. aeruginosa.* As in previous studies with TMV-CP, a cDNA was engineered so that the 9–14 mer peptide was to be inserted between the Ser154 and Gly155 of the CP. The cDNA of the chimeric TMV was transcribed *in vitro*. The resulting TMV RNA was used to infect tobacco cv Xanthi. The respective TMV-9–14 and wildtype TMV were purified from systemically-infected tobacco leaves.

The infection with TMV-9–14 caused the same symptoms and at about the same time after infection as wildtype TMV. The TMV-9–14 virus purified from the tobacco leaves reacted with anti-9–14 antibody. Electron microscopic analyses showed that the morphology of the TMV-9–14 was as of TMV, and the presence of the 9–14 mer epitopes was verified.

Immunization of five-week-old mice was performed according to standard procedures by the injection of 50 μg or 10 μg virus particles for several times with 14 days intervals [with $Al(OH)_3$, rather than with Freund's adjuvant, to simulate potential future human immunizations]. The expected titers of IgG were revealed in the immunized mice. It was found that, indeed, TMV-9–14 elicited antibodies reactive with peptide 9–14 mer purified from OM protein F. The immunoprotective potential of TMV-9–14 was analyzed in mice with chronic pulmonary infection caused by *P. aeruginosa.* Thus, mice were first immunized with either TMV wildtype or with TMV-9–14. These mice were then challenged with 5×10^3 agar-encased *P. aeruginosa,* delivered via intratracheal incision into the left lung of each mouse. The lungs were then analyzed by various criteria and it was found that the TMV-9–14 immunization caused a substantial (and statistically significant) reduction in lesions and retained bacteria in the lungs, as compared with TMV immunization. The authors rightfully conclude that their TMV model appears as an excellent basis for the development of a chimeric vaccine to protect against infections caused by *P. aeruginosa*.

Chapter 5

Therapeutic Products Unrelated to the Immune System

5.1. Introduction

This chapter will be divided into sections, according to the procedures of genetic transformation. These procedures include transformation by cauliflower mosaic virus (CaMV), infection by tobacco mosaic virus (TMV), transformation of plant-cell suspensions, regular *Agrobacterium*-mediated transformation, biolistics, and transformation by *Agrobacterium rhizogenes*. In the transformation by *A. rhizogenes*, the therapeutic product is formed in the roots. Thus, the product may be derived from roots of intact transgenic plants or, more commonly, the transgenic roots may be cultured *in vitro* and the product harvested from root cultures.

5.2. Transformation by Cauliflower Mosaic Virus (CaMV)

As was noted in the previous chapter (Sec. 4.3), investigators at the Friederich Miescher Institute in Basel attempted to employ CaMV as a vehicle to transform plants. One of the successful attempts was by de Zoeten *et al.* (1989). CaMV has a genome that is composed of a double-stranded circular DNA and is a retrovirus-related plant virus. It contains an open reading frame (ORF II) that can be replaced without impairing the CaMV infectivity. Replacing ORF II with a transgene should thus lead to the expression of the transgene in infected plants. CaMV will infect plants of the Cruciferae family, especially of the genus *Brassica*. The authors used a plasmid pCa35S in which most of the dispensable ORF II was replaced with a linker that allowed the insertion of an alien coding sequence. The investigators thus inserted the coding sequence for the human interferon (IFN-αD) or a deletion mutation of this coding sequence. The viral DNA of the authentic CaMV as well as the two modified CaMVs were cloned to result in Ca35S, Ca524^{i} and Ca562^{i} (Fig. 19). All the three were used to infect turnip plants (*Brassica rapa*). Both Ca524^{i} and Ca562^{i} induced the production of active IFN-αD in the infected plants. The virus

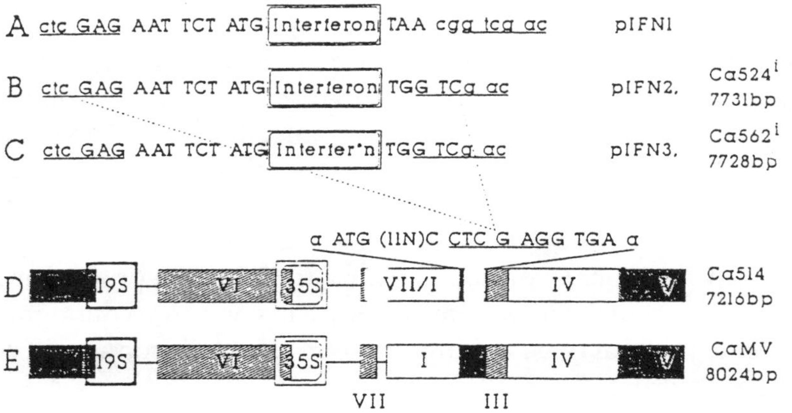

Fig. 19. Pertinent sequences of plasmids and constructs used in the manufacture of CaMV strains carrying the interferon gene. (A)–(C) Interferon-αD DNA sequences. Capital letters symbolize participation of nucleotides in the coding regions in the final constructs. The relevant restriction sites are underlined. The removal of codon 132 (TGT) in pIFN3 is indicated by *. (D) and (E) Comparison of CaMV with its strain Ca514. Ca514 cloned into pUC8 at its *Sal*I site is called pCa514 (not shown). The fusion/deletion in the ORF VII/I region and the deletion of ORF II are indicated. The remaining nucleotides of ORF II including the *Xho*I cloning site (underlined) are shown. The combinations of interferon fragments from the pIFN vectors with pCA514 are indicated, and resulted, after appropriate excision, in strains Ca524^{i} and Ca562^{i}, respectively. (From de Zoeten *et al.*, 1989.)

spread systemically after inoculation with both DNA clones. However, the spread was somewhat slower than after inoculation with the unmodified CaMV. Fully-assembled virus was revealed by electron microscopy. *In situ* localization of IFN-αD showed that the interferon accumulated in the inclusion bodies. The investigators also co-inoculated the turnip plants with the turnip yellow mosaic virus (TYMV) to test the possibility of the human interferon retarding plant viruses in plants. No such retardation was revealed and the authors were thus sceptic about a previous claim that human interferon can protect plants against pathogenic plant viruses.

5.3. Transformation by Modified Tobacco Mosaic Virus (TMV)

5.3.1. *Angiotensin-I-converting enzyme inhibitor*

The use of regular TMV for the expression, in infected plants, of an alien gene is problematic. Any impairment of the authentic coat protein (CP) will prevent the systemic spread of TMV. However, Hamamoto *et al.* (1993) found a way to add an alien gene to TMV and express this transgene in plants without any impairment of the CP. They used the effect of a six-base sequence (CAATTA) that was inserted immediately downstream of a stop codon triplet. When such an insertion is performed, there is a partial readthrough of the sequence downstream of the six-base sequence. The investigators thus intended to express in the plants the angiotensin-I-converting enzyme inhibitor (ACEI).

ACEI is a 12-amino-acid peptide (Phe–Phe–Val–Ala–Pro–Phe–Pro–Glu–Val–Phe–Gly–Lys) that is found in the tryptic digest of milk casein and it has anti-hypertensive effects when administered orally. The authors engineered DNA plasmids that code for TMV RNA by introducing the six bases 3′ of the CP code, added the triplet AGA (for Arg, to enable release of the ACEI by tryptic digestion) and then the code for ACEI. The engineered plasmids were transcribed and the transcripts were reconstituted with the TMV-CP. The reconstituted virus was used to infect leaves of young tobacco plants and leaves of tomato plants that had already borne some fruits. In tobacco, the leaves above the infected ones already produced TMV-CP and CP–ACEI. Infected tobacco plants maintained the production of CP and CP–ACEI for three months. In tomato, the virus moved from the infected leaves to uninfected leaves and the ACEI was also detected in the fruits (by immunoblotting with anti-ACEI antibodies). The leaves of the infected tobacco and tomato plants were estimated to contain about 100 μg CP–ACEI per g of fresh tissue. In tomato fruits, the estimate was 10 μg per g of fresh tissue. We saw no continuation of this study and are left with the intriguing question: Could chewing of these modified TMV-infected tobacco leaves (that contain ACEI) replace the soothing effect of milk?

5.3.2. *Alpha-trichosanthin*

Trichosanthin is a eukaryotic ribosomal-inactivating protein found in the roots of a Chinese medicinal plant. In *Trichosanthes kirilowii*, there is α-trichosanthin that cleaves an N-glycosidic bond in the 28S rRNA. This cleavage inhibits protein synthesis by affecting the ability of the 60S ribosomal subunit to interact with elongation factors. The mature protein is the result of several processing stages. Alpha-trichosanthin was considered as an anti-HIV drug and thus Kumagai *et al.* (1993) attempted to express this protein in plants inoculated with modified TMV.

As described in Sec. 4.3 (for antigenic epitopes on the coat proteins of plant viruses), a modified TMV is useful for expression of alien genes. In this modified TMV, the code for the endogenous CP is replaced by a code for the CP of another plant virus (odontoglossum ringspot virus, ORSV). The coding sequence for the transprotein is placed 3′ of the code for the 30 kDa protein ("movement protein") and upstream of the coding region for the CP of ORSV. The respective plasmid (pBGC152) was engineered by Kumagai *et al.* (1993) to express the transprotein α-trichosanthin. Therefore, the respective coding sequence for α-trichosanthin was placed under the control of the TMV CP subgenomic promoter (Fig. 20). Leaves of *Nicotiana benthamiana*

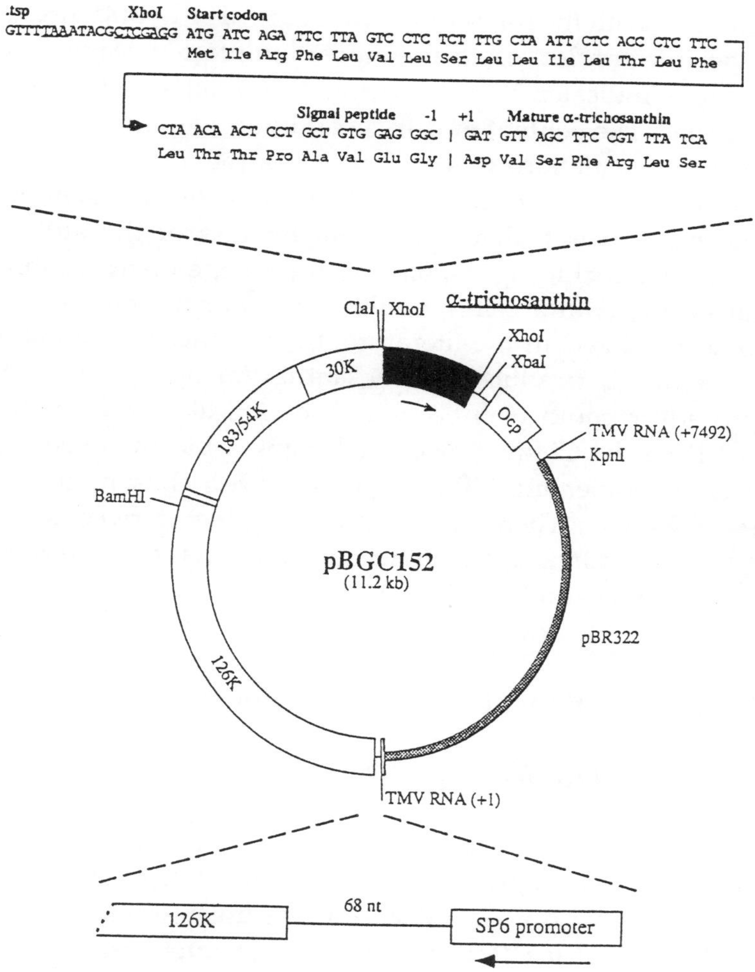

Fig. 20. α-*trichosanthin* expression vector pBGC152. This plasmid contains the TMV-U1 126- 183- and 30-kDa ORFs, the ORSV coat protein gene (*Ocp*), the SP6 promoter, the *α-trichosanthin* gene, and part of the pBR322 plasmid. The TAA stop codon in the 30 kDa ORF is underlined and a bar (|) divides the putative signal peptide from the mature peptide. The TMV-U1 subgenomic promoter located within the minus strand of the 30 kDa ORF controls the expression of α-trichosanthin. The putative transcription start point (tsp) of the subgenomic RNA is indicated with a period (.). (From Kumagai *et al.*, 1993.)

were infected with the transcript of pBGC152. The hybrid virus spread from the infected leaves to the upper uninfected leaves. Electron micrographs indicated that the virions were apparently identical to TMV virions. The content of α-trichoanthin in the upper leaves (two weeks after inoculation) was at least 2% of the total soluble protein. This is considered a rather high level. The α-trichosanthin from the infected leaves had all the characteristics of the authentic α-trichosanthin, including a concentration-dependent inhibition of protein synthesis in a cell-free rabbit reticulocyte translation assay. We could not trace, in the literature, the commercial exploitation of this procedure to produce α-trichosanthin for therapeutic purposes. The procedure could well be more economical than extracting this product from the Chinese medicinal plant — but then such projects are risky. If another anti-HIV drug replaces α-trichosanthin, the latter becomes obsolete. There are only a few plant-derived anti-cancer drugs that maintain a high commercial value over the years (e.g. taxol, vinblastine, vincristine).

5.4. Transformation of Cell Suspensions

5.4.1. *Human interferon*

Several years after de Zoeten *et al.* (1989) reported that human interferon-αD was expressed in transgenic turnip plants following transformation with a CaMV vector, a similar transformation was reported by Zhu *et al.* (1993). The latter investigators used a completely different transformation procedure. Their targets were rice-cell suspension cultures and rice plants. The main features of the procedure was to introduce a plasmid into the protoplasts of rice cells, culture these cells in a selective medium and then obtain transgenic cell suspensions and calli that produce human interferon-α. Calli that produced this interferon could be regenerated into rice plants.

The production of the primary cell suspension of rice is a time-consuming process (several months). It starts from the culture of young rice inflorescences in specific media, resulting in embryogenic calli that are then cultured in liquid medium with frequent (every ten days) subculture. Finally, a culture of small uniform clumps of cytoplasmically dense cells is obtained. The necessary enzymatic treatment converts the cells into protoplasts. The plasmid was introduced into the protoplasts with a "Lipofectin" reagent. The plasmid contained two cassettes. One cassette was for a selectable marker which caused the expression of the *npt*II gene (for resistances to kanamycin and neomycin). The other cassette contained the gene for human interferon-α with the necessary flanking promoter and terminator. The protoplasts were first cultured in semi-solid media and transferred to a selective medium for several weeks. The resulting colonies were either further cultured in liquid medium or transferred to plant regeneration conditions. Northern blot hybridizations indicated that a transcript for human interferon-α was produced in three transgenic rice plants. Bioassays on human amniotic WISH cells and infection with animal vascular stomatitis virus indicated that interferon activity existed in transgenic cultured cells and in transgenic rice plants. Since we are not aware of further reports on the production of interferon in transgenic rice, this endeavor was probably not pursued much further.

5.4.2. *Human erythropoietin*

Human erythropoieitin (Epo) is a hematopoietic growth factor involved in the regulation and maintenance of a physiological level of circulating erythrocytes. Epo is a highly glycosylated protein. The mature human Epo has three N-linked oligosaccharides at three amino acid positions (24, 38 and 83) of its 166 amino acid residues. The oligosaccharides are not essential for the *in vitro* activity of Epo but they are important for its stability in blood circulation. The human Epo has therapeutic use in certain cases of anemia. A group of Japanese researchers (Matsumoto *et al.*, 1995), all from the Kyoto area, investigated the production of human Epo in tobacco cell suspensions. The transformation procedure of these researchers differed considerably from the procedure described above (by Zhu *et al.*, 1993). An established tobacco cell line (BY-2) was used in this study. This line has a high cell division rate in suspension culture but it lost the capacity to regenerate tobacco plants. The transformation was performed by co-cultivation of BY-2 cells with an *Agrobacterium tumefaciens* line (LBA 4404) that harbored a helper plasmid and a binary plasmid (pBINCEP). pBINCEP had the LB and RB of the *Agrobacterium* plasmid T-DNA. Within these borders, there were two cassettes, one for the expression of *npt*II and one for the expression of Epo. The latter cassette contained a CaMV promoter, followed by a code for a 27 amino-acid signal peptide and behind the code for this signal, the coding sequence for human Epo was inserted. At the 3′ of this sequence, there was a *Nos* terminator. After a period of co-culture of the BY-2 cells and the agrobacteria, the bacteria were eliminated by antibiotics and the cells were further cultured in the presence of kanamycin. Colonies of kanamycin-resistant BY-2 cells were obtained and were screened for the production of Epo by ELISA, using monoclonal antibodies raised against recombinant human Epo. A single callus was (luckily) found (C3•5•2) that was a putative producer of human Epo. This apparent Epo was not released into the culture medium but was rather maintained in the tobacco cells. These authors established a large-scale production of cell suspension from callus C3•5•2 and found in 6.3 kg of cells the equivalent of 13 μg of Epo. The plant Epo was active in an *in vitro* assay based on measuring the erythroid colony formation of fetal mouse liver cells. The activity was even higher than that of authentic human Epo. But no activity was found in the tobacco-cell-derived Epo in *in vivo* assays. The authors attributed this lack of activity to the very different glycosylation of the Epo produced in tobacco cells as compared with the glycosylation of authentic human Epo. While the Epo was retained in the intact tobacco cells, it was readily released when these cells were converted into protoplasts. The authors suggested that the tobacco-cell-derived Epo may be useful as a growth factor in the *in vitro* propagation of erythrocytes.

5.4.3. *Human interleukin-2 and interleukin-4*

The lymphokines interleukin-2 (IL-2) and interleukin-4 (IL-4) are ligands for membrane-bound receptors. Both IL-2 and IL-4 are synthesized as polypeptides of 153 amino acids. These polypeptides contain signal peptides of 20 or 24 amino acids (in IL-2 and IL-4, respectively) that are cleaved off to result in the respective mature proteins. IL-2 mature protein is glycosylated and is normally synthesized and secreted by T-helper cells; it is thus an important component of the immune system. IL-4 is also an important component of the immune system. It exerts multiple immunomodulatory functions on a variety of cell types (e.g. T- and B–cells, and macrophages). A multinational team of researchers (Magnuson *et al.*, 1998) studied the expression of IL-2 and IL-4 in transgenic tobacco cells in suspension cultures. The investigators engineered binary plasmids containing the coding sequence for the 153 amino acids of IL-2 or the 153 amino acids of IL-4. In both cases, base changes were introduced upstream of the translation initiation codon (ATG) to conform with the usual plant sequences. The codes for interleukins were flanked with the 35S CaMV promoter and the T_7 terminator. The binary plasmids also contained a cassette for *npt*II expression. The two plasmids were termed pGA1560 (for the plasmid containing the code for IL-2) and pGA1361 (for IL-4). They were moved into the *Agrobacterium tumefaciens* line LBA 4404. Suspensions of agrobacteria containing either pGA1560 or pGA1361 were added to liquid cultures and co-cultivated with tobacco cells. After three days, the agrobacteria were washed off and further eliminated by the addition of cefotaxime. Selection in kanamycin-containing medium during several transfers to fresh selective medium resulted in kanamycin-resistant calli. Over 200 transformed calli resulted from the plasmid pG1560 (IL-2). These were maintained as calli. There was great variability in the levels of IL-2, as evaluated by ELISA. The highest levels were over 250 ng or over 350 ng IL-2 per g of cells. When put into liquid culture, the levels of IL-2 peaked five days after transfer and most of the IL-2 was retained in the cells with only about 1/100 released into the medium. Cells transformed with agrobacteria containing pG1361 (IL-4) also yielded over 200 kanamycin-resistant calli. The IL-4 yields were much higher than the yields of IL-2. Also, when cultured in the liquid medium, the IL-4 yields were higher and the levels of IL-4 released into the medium approached the levels of IL-4 in the cells (about 180 and 275 ng per ml of culture, respectively).

Western blot hybridization using anti-IL-4 antibody indicated that the IL-4 in the tobacco cells and in the medium was similar to authentic (commercial) IL-4. When the cells that produced IL-2 were analyzed for the biological activity of their IL-2, it was found that only the medium contained biologically active IL-2. The same was revealed for cell suspensions that produced IL-4: the biological activity was found only in the medium but not intracellularly. However, the biological activity in the medium was lower than expected from the levels of IL-4 revealed by ELISA. It should be noted that the plasmids did not contain signals to retain the interleukins in the ER nor to specifically cause their release from the cells. The mode of processing and glycosylation of the interleukins in the plant cells also require further investigations. Hence, the authors suggested further studies to render the manufacture of human interleukins in transgenic plant cells a feasible biotechnology.

5.5. *Agrobacterium tumefaciens*-Mediated Transformation

In this section, we shall review various therapeutic products that were expressed in transgenic plants. The common denominator for these products is that they were all expressed in transgenic plants that were obtained after conventional *Agrobacterium*-mediated transformation.

5.5.1. *Leu-enkephalin*

Leu-enkephalin is a pentapeptide (Tyr–Gly–Gly–Phe–Leu) located in the brain and it has opiatic activity. A group of investigators from Gent, Belgium (Vandekerckhove *et al.*, 1989) chose this pentapeptide as a model system to express therapeutic peptides in transgenic plants. Chronologically, this was a pioneering effort in this direction. It is noteworthy that in this study several aspects were considered in the planning of the research that were ignored in later studies by other investigators. For example, Vandekerckhove *et al.* (1989) aimed the expression of the transgene to an organ where it can be accumulated in large quantities — the protein bodies of seeds. They also constructed the cassette of the transgene in a way that will permit easy separation of the resulted leu-enkephalin from the seed protein.

The authors chose the 2S albumin of *Arabidopsis* seeds and inserted the code for the alien pentapeptide in a variable domain of a 2S albumin in the seeds of this plant. They avoided any changes in the 2S albumin that would change the secondary structure of this protein. The insertion was thus performed between the 6th and the 7th cysteine. They also flanked the code for the five amino acids by codons for lysine. This was done to enable tryptic digestion of the 2S albumin and release of the pentapeptide from the digest. A respective transformation vector (PGSATE1) for *Agrobacterium* transformation was constructed. The cassette for selection contained two genes for resistance to antibiotics; they usually used *npt*II gene and a gene for resistance to hygromycin (*hgh*). The cassette for 2S albumin expression was also flanked by a promoter and a terminator but the sequence was changed to code for the following amino acids:

original sequence: ...CPTLKQAAKAVRLQGQHQP...
modified sequence: ...CPTLKQAA<u>K</u><u>YGGFL</u><u>K</u>QHQP...

In these sequences, the lysines for trypsin digestion are double-underlined and the leu-enkephaline is underlined.

Because the N-terminal and the C-terminal ends of the 2S albumin were not changed, the protein was regularly trafficked into the seeds and the transpeptide accumulated in protein bodies. Using the above mentioned transformation vectors, the authors first transformed *Arabidopsis thaliana* plants. They used the "old" method of incubating leaf pieces with *Agrobacterium* suspensions as the current procedure to transform *Arabidopsis* (see Sec. 2.4) was not yet available. Five transgenic *Arabidopsis* plants that produced up to 200 nmol pentapepticle per g of seeds, were obtained. In further studies, the authors transformed *Brassica napus* (oilseed rape). The highest level of leu-enkephalin in the seeds of this plant was 50 nmol per g of seeds. The seed yield of rape is rather high, about 3,000 kg per hectare. Thus, the authors went into the optimistic calculation that a hectare of transgenic rape should yield up to 75 kg of leu-enkephalin. Moreover, the extrapolation went further: as there are 12 or more 2S albumins in rape and the pentapeptide could be inserted into all of these, the yield could be vastly increased. Does the medical world need 1,000 kg of leu-enkephalin and what about co-suppression when 12 transgenes are integrated into one genome? With a smile, we could only imagine what a "high" feeling cows will be having when they happen to find rapemeal (the material left after oil extraction) with leu-enkephalin in their feed (bumble bees easily transfer rape pollen from one field to another).

5.5.2. *Human serum albumin*

Human serum albumin (HSA) is initially synthesized in the liver as a pre-pro-albumin. The latter is released from the ER after removal of 18 amino-terminal amino acids, resulting in the respective pro-albumin. The pro-albumin is additionally processed in the Golgi complex where a further six amino acids are cleaved by a serine proteinase. This results in the secretion of the mature HSA. Previous attempts to express HSA in either bacteria or yeast did not result in the production of mature HSA. A group of investigators of the Mogen International Company in Leiden, The Netherlands (Sijmons *et al.*, 1990) showed that mature HSA, albeit in very minute quantities, can be produced in the leaves and stem tips of transgenic potato plants. Their rationale was that the human signals may lead to correct trafficking in the potato tissue and that peptidases of the transgenic plant may recognize the pre-pro sequences of the HSA (or a modified pre-pro sequence) and these would be cleaved by the plant enzymes. These investigators used the (now) conventional protocol of *Agrobacterium*-mediated transformation by the binary vector system. Their binary HSA expression vectors contained a cassette for resistance to kanamycin as well as an HSA cassette. These cassettes were flanked with the T-DNA LB and RB. As for the HSA cassette, they engineered several versions of the code for the mature HSA and its pre-pro sequence. In all cases, they used the 35S CaMV promoter with some modifications (with a duplicated enhancer and a synthetic leader sequence derived form the alfalfa mosaic virus RNA). The *Nos* terminator was fused downstream of the HSA coding sequence. In one construct (pMOG236), the authentic code for the pre-pro sequence was retained; two other constructs contained a fusion of the mature HSA with the pre-sequence from the extracellular pathogen-related S (PR-S) protein. In one of these two constructs, the fusion was between the code for the C-terminal of the PR-S and the code for the N-terminal of the HSA (pMOG250). In the other construct (pMOG249), an extra codon was added between these two codes. As control (pMOG285), they fused directly the code for the mature protein to the promoter (with leader). Transgenic potato plants were obtained and analyzed for the respective transcripts. Most plants that were analyzed were positive. Plants transformed with pMOG250 yielded mature HSA with the correct N-terminal amino acid (Asp). The plants transformed with pMOG249 also produced the mature HSA but the latter was preceded with an Ala. The mature HSA in the transgenic plants was secreted to the intercellular spaces. Also, transgenic suspension cultures secreted the mature HSA into the culture medium. The yields of the mature HSA were low: 0.02% of total soluble protein in leaves and 0.25 μg HSA per mg of protein in the culture medium. The authors suggested that much higher yields could be obtained in the future in potato tubers if the class I patatin promoter rather than the modified 35 S CaMV promoter would be used. Their extrapolation was far reaching: if the transgenic tubers would contain 1–2% HSA, then a potato field of one acre (50 tons of tubers) should yield 12 kg of HSA (starch, in the authors' opinion, would be an extra benefit). We did not see yet further reports in the reviewed literature on potato fields that produce HSA.

5.5.3. *Cytochrome P-450*

The P-450 proteins are protoheme-containing monooxygenases involved in the oxidative metabolism of a broad range of substrates. These proteins were found in all eukaryotic organisms (e.g. fungi, angiosperm plants and mammals). In plants, P-450 enzymes are involved in oxidative reactions in the biosynthetic pathways of secondary metabolites and in the detoxification of herbicides. In mammals, P-450 enzymes produced in the liver have an important role in xenobiotic and drug metabolism. Thus, cytochrome P-450 of mammalian liver is interesting for basic mammalian metabolic studies and for biomedical studies. A group of Japanese investigators researched on the possibility of producing cytochrome P-450 in transgenic tobacco plants (Saito *et al.*, 1991). These investigators constructed a binary vector (pSN002) that contained two cassettes within the RB and the LB of *A. tumefaciens* T-DNA. One was the common *npt*II cassette for kanamycin resistance. The other cassette contained the cDNA for cytochrome P-450 of rabbit liver (after induction by phenobarbital). This cDNA was flanked by a mannopine synthase promoter and the termination polyadenylation sequence of gene 7 from the T-DNA of *Agrobacterium*.

Agrobacterium-mediated transformation was performed in the conventional manner by incubating tobacco leaf discs in a suspension of agrobacteria that contained the pSN002 vector as well as a helper plasmid. The bacteria were then eliminated by washing and antibiotic treatment, and selection of transformed calli and shoots was done in the presence of kanamycin. The authors observed that transformation with a cDNA for P-450 took longer than with other transgenes; about six months were required to obtain the putative transgenic plants. The authors' suggestion was that this specific transgene caused the delay. Southern blot hybridization showed that one to three integrations of the transgene occurred per tobacco genome. Northern blot hybridization indicated that the transgene was transcribed to different levels in the individual transgenic plants. Western blot analysis indicated low (but significant) levels of transgenic P-450 in the microsomal fraction.

Interestingly, the alien P-450 had a profound effect on the morphology of the transgenic plants: they had a tendency to early senescence and flowering. They also had a very high level of malondialdehyde and their alkaloid metabolism was changed. The P-450 was apparently accumulating in the phloem of the transgenic plants. The take-home lesson is not that transgenic plants can serve as a good source of P-450, but rather that an overdose of a specific enzyme that has a parallel in plants may be detrimental on plant growth and metabolism. Therefore, transgenic plants cannot serve as a good source of such alien enzymes.

5.5.4. *Human epidermal growth factor*

Higo *et al.* (1993) in Japan surveyed the problems of expressing mammalian genes in transgenic plants, that were attempted in previous studies. One of the problems, they considered, was the correct processing of proteins that are translated as precursors and undergo changes, commonly N- and C-terminal cleavages. To overcome this problem with respect to the gene that encodes the human epidermal growth factor (hEGF), they decided to use the synthetic cDNA for the mature hEGF. They obtained such a sequence in which the codons for amino acids fitted translation in bacteria (*Escherichia coli*). At that time (1993), the authors were not aware of the subsequently accumulated information that the codon choice of bacteria is very different from that in plants. Hence, the sequence they chose could substantially retard the protein synthesis in plants. This problem was faced by the investigators at the Monsanto Company (and affiliated laboratories) when they attempted to express in transgenic plants the bacterial Bt for insect toxicity. Anyway, the mature hEGF is a small (M_r 6,000) but powerful mitogenic protein. The authors changed a few bases 5′ of the synthetic code for hEGF to conform to the common plant sequence upstream of the translational initiation. They flanked this modified code with the 35S CaMV promoter and terminator and inserted it into a binary vector that also contained the *npt*II gene for kanamycin resistance. They then performed the conventional *Agrobacterium*-mediated transformation. This resulted in six tobacco plants that were resistant to kanamycin. In five of them, they found (by Southern blot hybridization) the integrations (one or more integrations per genome) of the hEGF gene. At least some of the progeny plants, after self-pollination of the originally transformed tobacco plants, showed the production of hEGF. One plant, T-7, and its progeny were analyzed further for the production of hEGF by ELISA. This test was supposed to have a sensitivity to detect 50 μg per ml, or higher levels of hEGF. Only a single plant of the T-7 progeny had apparently 65 μg/ml; all the others had less than 40 μg/ml of leaf extract. That means that all the progeny plants but one had levels of hEGF that were not significantly above the control tobacco plants. It was probably due to these low levels of hEGF, as revealed by ELISA, that no bioassays were performed. By hindsight, these results are not surprising, because on the basis of several subsequent studies, we know that the expression of alien protein in the cytosol will at best result in very low yields.

5.5.5. *Trout growth hormone*

A team of investigators from two Belgium biotech companies (Plant Genetic Systems and Eurogentec) followed the expression, folding and glycosylation of the growth hormone of the rainbow trout (Bosch *et al.*, 1994). Before we discuss this study, the story of Aldous Huxley,

"After Many a Summer Dies the Swan", comes to mind. The essence of this story is that a juvenility factor was revealed in carps (carps are known to grow continuously for more than 100 years) by an English couple. This factor was so effective that the couple, who fed on it, were rended so "juvenile" that they were transformed into apes.... Thus, with the help of Huxley's story we shall include the trout growth hormone in our deliberations on therapeutics.

As Higo *et al.* (1993) mentioned above, Bosch *et al.* (1994) also intended to investigate the fate of alien proteins that were expressed in transgenic plants. Will such proteins be stable in the plants? Will processing from precursor to mature protein take place? What about correct folding and glyocosylation?

The trout growth hormone (tGH-II) undergoes several processings in trouts. There is an N-terminal signal sequence that is co-translationally removed during passage into the ER; two disulfide bridges are formed and there are two potential glycosylation sites on the protein that contains 188 amino acids. For *Agrobacterium*-mediated transformation, they used binary vectors that also contained a selective gene cassette. Two vector plasmids were constructed for the transformation of tobacco. In both, the code for the mature tGH-II (with minor modifications) was driven by the promoter for the small subunit of Rubisco. In one of these, the investigators added a code for the signal peptide of the PR1-b for the secretion of the transprotein into the intercellular spaces. Also, two vector plasmids were constructed for the transformation of *Arabidopsis.* In both, they inserted the Pat2s1 promoter that causes expression in the seeds. In one of these, they also added the code for the signal peptide of PR1-b.

RNA extracts from leaves of transformed tobacco plants and from seeds of transformed *Arabidopsis* plants were used for Northern blot analyses. These analyses confirmed that the transcripts for tGH-II were synthesized in the transgenic plants. When the tGH-II protein was analyzed, the correctly processed and folded protein was found only in leaves of tobacco transformed with a vector that also contained the code for the PR1-b signal peptide. When the vector lacked this code, causing cytoplasmic expression, there was misfolding and partial degradation of the tGH-II. The final product as evaluated by radioimmunoassay was rather low: 43 ng/ml of partially purified leaf extract samples, this was only 1% of what was estimated from immunoblots (0.1% of soluble protein). This suggested to the authors that most of the tGH-II protein in the tobacco leaves "is in a form unlikely to be biologically active". It could be that the extraction and purification changed the final product. The authors terminated their discussion by stating that: "the expression of a foreign protein in plants requires careful study of the parameters other than just promoter strength, including mRNA stability and differential protein stability in different cellular environments." We agree.

5.5.6. *Ricin*

When the prophet Jonah came close to the city of Nineveh (Hebrew Bible, Jonah, Chap. 4), he took shelter under a castor plant (*Ricinus communis*). Luckily (for Jonah), he was so tired that he did not taste the fruits and went to sleep. When he woke up, there was no castor plant anymore. If Jonah would have eaten the fruits, he rather than the castor plant would have died: castor beans contain ricin, one of the most potent plant-derived cytotoxic proteins. Because of its extremely high toxicity, ricin was considered as an agent for cell-targeted ablation (i.e. when attached to another agent that would lead ricin specifically to the targeted cell). Ricin belongs to a family of ribosome-inactivating proteins (RIP). It is a heterodimer consisting of two chains: an RNA-specific N-glycosidase *a*-chain (RTA) that is disulfide-linked to a galactose-binding lectin *b*-chain (RTB). The toxin is synthesized in castor seeds ("castor beans") as a 576-amino-acid pre-pro-ricin. It has initially a 35-amino-acid N-terminal signal sequence and a 12-amino-acid linker sequence, between the RTA and RTB. It is assumed that the signal sequence leads the pre-pro-ricin into the ER where the signal is removed. From the ER, the pro-ricin is moved to the protein storage vacuoles of the seed where the final mature, two-chain ricin is accumulating. The production of ricin in animal systems is problematic because of inaccurate processing and high toxicity to animal ribosomes. Plant ribosomes are much less sensitive to ricin than animal ribosomes. Sehnke *et al.* (1994) thus attempted to express the ricin gene in transgenic tobacco plants. They used the existing cDNA for the pre-pro-ricin and engineered it into a binary vector for *Agrobacterium*-mediated transformation. This cDNA was flanked by the 35S CaMV promoter and a terminator. The engineered vector was moved into the *A. tumefaciens* strain LBA4404. Regular co-cultivation of tobacco leaf discs with the agrobacteria was followed by elimination of the bacteria and culture in selective medium (kanamycin). The authors did not detail how many kanamycin-resistant plants were obtained. Protein extracts of leaves from the transgenic tobacco plants were used in immunoblot analyses with either anti-α-RTA or anti-α-RTB antibodies. These analyses indicated the presence of RTA and RTB in the plant leaves. Moreover, the analyses also suggested that only the processed heterodimer accumulated in the tobacco leaves and no unprocessed pre-pro-ricin was produced. The latter situation is very relevant because if the RTA should be used for therapeutic purposes, the product should be free of pre-pro-ricin contamination. The level of ricin was evaluated by ELISA and estimated to be 1 μg of rec-ricin per g of fresh weight of leaves. This level is about 0.25% of the soluble proteins of the leaves. The activity of the rec-ricin was analyzed by the endoglycosidase activity test that measures protein translational inhibition in rabbit reticulocyte lysate. There was a high activity of the rec-ricin: 50% inhibition by a dose of 3×10^{-11} M. Also direct cytotoxicity was tested. This test combines the binding capability of rec-ricin as well as its cytotoxicity. It was found that for HUT102 cells the ID_{50} was 1×10^{-12} M.

Five years after the original publication of this system by Sehnke *et al.* (1994), Sehnke and Ferl (1999) provided a more detailed account on the processing of the pre-pro-ricin in transgenic tobacco. The original results were thus confirmed. For example, 200 g of transgenic tobacco leaves yielded 200 μg of soluble rec-ricin. The separation of this rec-ricin on SDS–polyacrylamide gel showed the expected bands for the 30–35 kDa chains and the bands hybridized specifically with the anti-α-RTA and anti-α-RTB antibodies, respectively. The bands were also sequenced (Edman degradation sequencing) confirming that the bands represented authentic RTA and RTB; although there was a minor component of a modified rec-RTB. The two rec-ricin chains could be separated after DTT treatment, as it happens with authentic ricin. The recent report also showed that the rec-ricin was produced in cell suspensions derived from the transgenic tobacco plants. In this case, the rec-ricin was also released into the culture medium reaching 25 ng/ml. Changing the level of plant hormones in the culture medium affected the level of rec-ricin production and the authors suggested that future optimization of plant hormone level could further increase the rec-ricin yields.

5.5.7. *Hirudin*

Hirudin is the most potent and specific thrombin inhibitor known. No wonder, the efficient feeding of leech (*Hirudo medicinalis)* depends on it. This is because thrombin activity needs to be inhibited when leeches feed on the blood of their hosts. This small protein (about 7 kDa) is produced in the salivary glands of the leech and it is being studied in clinical trials for use as a blood anticoagulant. The cDNA encoding hirudin is available. Hence, Parmenter *et al.* (1995) considered hirudin as a good model to study the manufacture of an alien protein in transgenic plants. These authors planned their study meticulously. First, they realized that the codon usage of the leech had to be changed to the codon usage of plants (e.g. rape and *Arabidopsis).* Then, the authors considered where to express the hirudin gene. The seeds of transgenic plants seemed to them an appropriate location but seed storage proteins can include only a very limited number of additional peptides. They therefore chose another "carrier" for the transprotein: the seed-specific oleosins. These are small but abundant proteins embedded in the phospholipid monolayer of oil bodies. In some seeds such as that of oilseed rape, they represent up to 20% of the total seed proteins. Moreover, they can be purified easily. So, attachment of hirudin to oleosin was an attractive possibility. But these investigators went further in their planning: after the isolation of oleosin from the seeds, the hirudin should be separated; thus, they intended to add a proteolytic cleavage site between the oleosin and the hirudin.

In practice, Parmenter *et al.* (1995) engineered an oleosin–hirudin fusion gene. For that, they produced a synthetic cDNA for hirudin, with the Cruciferae codon usage. They took the *Arabidopsis* promoter for oleosin as well as the cDNA for this oleosin, and between this coding sequence and the synthetic cDNA for hirudin, they inserted a sequence encoding a proteolytic cleavage site (Factor Xa/clostripain). At the end of this fusion gene, they put a *Nos* terminator. The fusion gene construct was inserted into an *Agrobacterium* strain (EHA) by electroporation. *Brassica napus* (rape) petioles were then transformed with these agrobacteria and after due selection on kanamycin, four transgenic rape plants were obtained. Total RNA of developing seeds was analyzed by Northern blot hybridization. Three out of the four putative transgenic plants had the hirudin transcript in their seeds. The isolation of oil-body protein enabled the analysis by immunoblotting that revealed the oleosin–hirudin fusion protein in the transgenic rapeseeds. The expected molecular mass of 26 kDa (19 kDa of oleosin plus 7 kDa of hirudin) was revealed. The existence of hirudin was also confirmed by immunofluorescence of the isolated oil bodies (Fig. 21). The activity of the hirudin produced in the transgenic plants was measured by a colorimetric thrombin inhibition assay. In this assay, the investigators tested wildtype oil bodies, with or without proteolysis (Factor Xa), as well as oil bodies from the transgenic seeds, with or without proteolysis. Antithrombin activity was revealed only in the transgenic oil bodies after proteolysis. These results clearly indicated that transgenic rapeseeds are a potential source of hirudin. The authors, as others before and after them, made

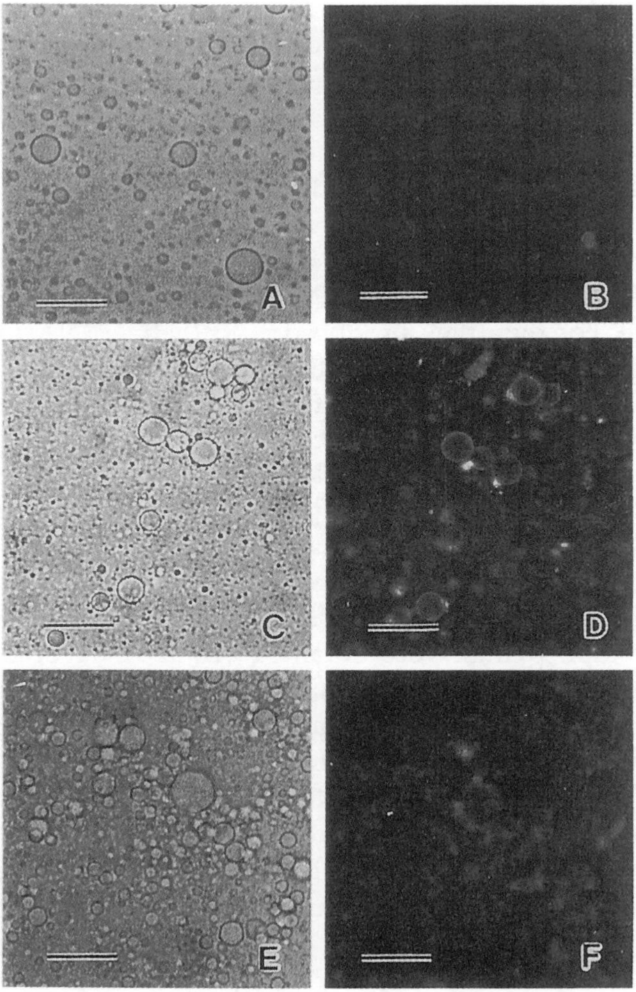

Fig. 21. Immunofluorescence localization of hirudin. Oil bodies from transformed and wildtype seeds were isolated and probed with anti-hirudin monoclonal antibodies and the appropriate FITC-conjugated secondary antibodies (A), (C) and (E): Light microscopy. (B), (D) and (F): Fluorescence microscopy. Transformed oil bodies were treated (E, F) or untreated (C, D) with Factor Xa. As shown, oil-body fluorescence is significantly reduced after Factor Xa treatment. Wildtype oil bodies (A) demonstrated no fluorescence (B). Scale bar = 25 μm. (From Parmenter *et al.*, 1995.)

financial calculations. They indicated that the cost of one ton of rapeseeds is US$250. The oleosin–hirudin fusion protein could, according to their estimate, reach 1% of the seed protein. From this calculation, the cost of 1 kg hirudin should be very low. This is a very naive calculation. Many expenses were not taken in account; for example, cleavage has to be performed with the very expensive Factor Xa. Nevertheless, this study is one of the best examples of the manufacture of a therapeutic product by transgenic plants.

5.5.8. *Human milk protein β-casein*

Human milk differs substantially from cow milk. The former contains 0.9% protein of which casein comprises 20%. The β-casein of human milk is the most abundant casein, having a molecular mass of 25 kDa. It has from none to five phosphate groups attached to Ser and Thr residues within the first ten amino acids of the N-terminal end of the peptide chain that facilitates binding to divalent metal ions. Cow milk β-casein is completely phosphorylated at the N-terminal end and also differs by additional respects from human β-casein. Since many years ago, there has been accumulated evidence that cow milk may cause severe infant maladies. In very rare cases, even human milk from mothers who consumed cow milk may cause an anaphylactic shock. In short, there is a good reason to substitute the β-casein of cow milk with recombinant protein of human β-casein in infant formulations. Thus, to construct plant-based infant formulas that resemble human milk, Chong *et al.* (1997) performed a research to express human β-casein in potato. Consequently, they engineered an expression vector that had the following components:

— an *npt*II cassette for kanamycin resistance;
— a P1,2 mannopine, two-directional promoter (*mas*);
— a bacterial luciferase gene (*lux*F) as reporter gene;
— the cDNA for the β-casein of human milk with the leader, flanked by the P2 of *mas* and a terminator sequence.

The expression vector was introduced into an *Agrobacterium* strain by electroporesis and the transformation was performed on the leaf explant of potato (c.v. Bintje). After due co-cultivation of the leaf explants with the agrobacteria, the latter were eliminated (by claforan) and the explants were cultured in kanamycin-containing medium. Regenerated plantlets that rooted on kanamycin-containing medium were transferred to the greenhouse. Putative transgenic potato plants were detected by low light image analysis (Argus-100 intensified video camera) that revealed light emission due to the expression of the *lux*F gene. Six putative transgenic potato plants were isolated. PCR reaction with these plant DNAs revealed that they were indeed transgenic and contained the human β-casein gene. Reverse transcription PCR (RT-PCR) revealed that the human β-casein transcript was produced in all the six transgenic potato plants. The recombinant β-casein protein was found in the extract of leaves from the transformed potato plants by immunoblotting with anti-human-β-casein antibody. On the SDS–PAGE gel, the β-casein from potato appeared to be 1–1.5 kDa smaller than the authentic β-casein. It should be noted that the *mas* promoter is induced by wounding and by auxin. Thus, the β-casein expression in leaves was detected after auxin application and wounding, and reached 0.01% of total soluble protein. Even lower levels of human β-casein were detected in the tubers of the transgenic plants. This casein was actually detected in tubers only after five days of induction with auxin. The authors expressed their intention to use other promoters in the future (e.g. the patatin promoter for expression in tubers). Nevertheless, the authors made calculations for the future. They reasoned that if 33,400 pounds of potato tubers were to be produced in one acre, and if 2% of the tuber proteins were to be human β-casein, this should provide 6.4 kg human β-casein per acre or the equivalent amount of β-casein in 7,076 liters of human milk. This optimistic calculation led the authors to make a statement: "The reconstitution of human milk in edible plants will ultimately lead to the development of infant formulas and solid baby foods with increased nutritional and therapeutic potential as well as increased protection of neonates against the development of food allergies and autoimmune diseases". We leave it to the readers to judge this statement.

5.5.9. *Human hemoglobin*

There is a need for hemoglobin-based blood substitutes. One source of such substitutes is outdated human blood. However, several problems, such as the presence of infectious agents, diverted the attention to new sources of hemoglobin. A team of eight investigators from the Hôpital de Bicetre (Le Kremlin-Bicetre, Paris) a site with a long tradition of hematological research (originally an asylum) and from "Limagrain" (a large plant breeding company) in Aubiere, France, looked to transgenic plants as a source of human hemoglobin (Marden *et al.*, 1998). Up to now, we have only a "scientific correspondence" report in the journal *Nature* and a summary of a lecture in 1998 from this team. However, the initial efforts look encouraging. These investigators "sent" the α- and the β-hemoglobin into tobacco plastids. They used the conventional *Agrobacterium*-mediated transformation system and an engineered binary plasmid to obtain transgenic tobacco plants. The plasmid contained a kanamycin-resistance cassette and another cassette that included the cDNAs for the human α- and β-globin that were fused downstream of the transit peptide of the small subunit of Rubisco from peas. This transit peptide should lead the globins into the chloroplasts (and the plastids). The proteins of transformed plants contained the recombinant hemoglobin (rHb) and the rHb was found to be functional. The level of rHb in the tissues of the transgenic tobacco plants was rather low but the authors have in mind to use other plants to increase the yields of rHb. Moreover, this procedure should permit specific changes in the cDNAs for Hb and thus, the oxygen affinity of the rHb could be fine-tuned.

5.5.10. *Human α-lactalbumin*

Alpha-lactalbumin (α-LA) has a major role in the production of lactose in milk. In the mammary gland, it combines with the Golgi-associated β-1,4-galactosyltransferase to form the lactose synthase complex. The primary and tertiary structures of α-LA are known and the cDNA of human α-LA (for the leader peptide and the mature protein) is available. Takebe and Hagiwara (1998) of Tsukuba, Japan, tested the production of human α-LA in transgenic tobacco plants. They engineered two binary vectors, one with the signal peptide coding sequence upstream of the code for the mature α-LA and the other with the coding sequence encoding the signal leader. They also flanked the cDNAs with a promoter (a modified 35S CaMV promoter) and a terminator. The vectors also contained a cassette for *npt*II. Conventional *Agrobacterium*-mediated transformation was performed with tobacco leaves and after selection on kanamycin-containing medium, several putative transgenic tobacco plants were obtained. PCR and RT-PCR analyses indicated that the α-LA gene was inserted in the plants genomes and also produced the expected transcripts. The production of α-LA was followed by Western immunoblotting. Only one plant (B2-2) which was transformed with the vector that included the coding sequence for the signal peptide provided a band that corresponded with α-LA. It seemed that the α-LA was correctly processed. The α-LA was partially purified from the leaves of B2-2 and assayed for its ability to promote lactose synthesis. The results were positive. The yield of α-LA in B2-2 was rather low (about 40 μg/ml of leaf extract) but the authors argued that several future modifications of the procedure could increase the level of α-LA considerably.

5.5.11. *Human lactoferrin*

Human lactoferrin (hLf) is an 80 kDa monomeric and bilobed glycoprotein. It was originally found in human milk but it was then detected in other epithelial and mucosal secretions as well as in neutrophils. The protein is able to bind tightly but reversibly to ferric ions. It is a multifunctional protein and has antibacterial, antifungal and antiviral activities. It is assumed that the antibacterial activity results from the ability of hLf to deprive bacteria of iron, but hLf may also have a more direct antimicrobial activity by changing the microbial cell membrane and causing lysis. In addition, there are various other functions attributed to hLf, some of these are correlated with the immune system. In short, it has potential therapeutic applications. There were publications on the production of recombinant hLf in cultured mammalian cells but there are obvious safety and other advantages of producing functional hLf in transgenic plants. Such a production was the goal of Salmon *et al.* (1998) — a team of several groups of investigators (11 in total) from France. These investigators engineered two binary vectors for *Agrobacterium*-mediated genetic transformation. In one vector, they retained the code for the native signal peptide while in the other vector they exchanged this code with one coding for the signal peptide of sweet potato sporamin. The two vectors also had all the other usual components (e.g. a cassette for kanamycin resistance) as shown in Fig. 22. The usual transformation procedure yielded 20 putative transgenic tobacco plants for each of the two vectors. They retained one transgenic tobacco plant (T19) that contained the human signal peptide attached to the hLf protein and one plant (T30) with the sporamin signal peptide. T19 and T30 contained 0.1% and 0.3%, respectively, of hLf in the soluble proteins in the leaves. Western blot analyses revealed in leaves of T30 a band that reacted with rabbit anti-milk-hLf primary antibody and this band migrated (on SDS–PAGE) as the authentic hLf, with an apparent molecular mass of about 80 kDa. Sequencing of the N-terminal amino acids of the recombinant hLf showed identity with the sequence of authentic hLf. Mass spectrometric analysis hinted that glycosylation of the recombinant plant-derived hLf differed from the glycosylation of authentic hLf. More specific analysis revealed differences in certain sugars of the two types of hLfs: the recombinant hLf had low galactose and contained xylose but lacked N-acetyl-neuraminic acid, while the

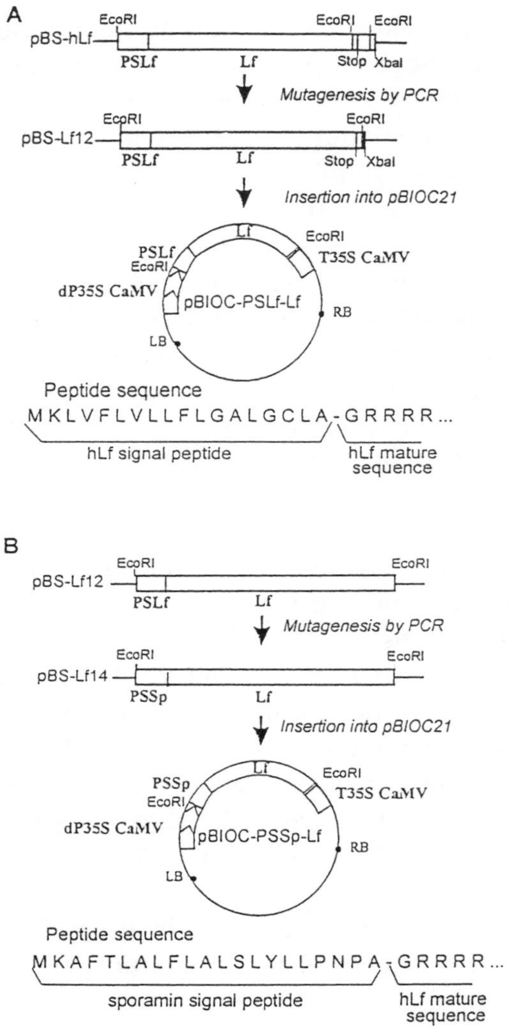

Fig. 22. Schematic representation of the binary plant expression vectors (A) pBIOC-PSLf-Lf and (B) pBIOC-PSSp-Lf. N-terminal amino acid sequences of the various fusion proteins are shown below the diagram. Abbreviations: PSLf, signal peptide of human lactoferrin; PSSp, signal peptide of sporamin; dP35SCaMV, enhanced 35S promoter of CaMV; T35S CaMV, 35 S termination signal; LB/RB, left/right T-DNA border sequence. (From Salmon *et al.*, 1998.)

authentic hLf had more galactose, no xylose and did contain N-acetylneuraminic acid. As previous authors before them, Salmon *et al.* (1998) expressed the optimism that future work possibly with additional agronomic species, such as "corn" (they mean *maize;* "corn" is an ill-defined term for cereals with a different meaning in different countries — usually the term "corn" is used for the most prevalent cereal in a given country), will enable large-scale manufacture of the recombinant protein in transgenic plants.

It should be noted that several years earlier, Mitra and Zhang (1994) introduced the hLf gene into tobacco cells grown in cell suspension culture. But these authors intended to render tobacco resistant to pathogenic bacteria and the transprotein in the tobacco cells was much smaller than authentic hLf.

5.5.12. *Human granulocyte/macrophage colony-stimulating factor*

Sardana *et al.* (1998), who referred to previous literature, estimated the initial capital expenditure for the production of proteins from transgenic sources. This expenditure for mammalian cell lines was estimated at US$25 to 50 million but only US$250,000–500,000 for stably transformed plants. We recommend taking these numbers "with a grain of salt" but agree that production of therapeutics in transgenic plants could be cheaper than their production in mammalian cell cultures. Based on their estimates, Sardana *et al.*, (1998) worked out a detailed laboratory protocol on how to express the human cytokine, granulocyte/macrophage colony-stimulating factor (GM-CSF) in the seeds of transgenic tobacco. While the procedures and plasmid construction strategy of this publication make good sense, the authors did not present experimental results. The general strategy recommended in this protocol is that in order to be expressed in seeds, a seed-specific promoter has to be used. They recommended the rice glutelin promoter *Gt3* but other seed endosperm promoters were noted as alternatives. A code for the signal peptide of glutelin to lead the transprotein to its destiny was also recommended, and this again should be followed by a code for a few amino acids of the N-terminal end of a seed protein (glutelin); downstream of that should come the coding sequence for the transprotein and finally a terminator sequence. We do not know how efficient this protocol really is. A full research article will tell us, hopefully.

5.5.13. *Human protein C*

Human protein C (hPC) is an important anticoagulant. hPC is highly processed and has to interact with other components to exert its activity. Before its secretion from the liver cells into the blood, the hPC zymogen has to undergo several modifications and the pre- and pro-signals have to be removed to yield light (M_r 21,000) and heavy (M_r 41,000) chains. The latter is further modified by the removal of a short (12 amino acids) peptide. There is no full research article on the production of hPC in transgenic plants but there are reports on such a study in two reviews (Cramer *et al.*, 1996; 1999). Attempts to express hPC in tobacco plants by using an expression vector with the 35S CaMV promoter (with modifications) were performed. At least some of the required processing and glycosylations were accomplished in the transgenic plants but the yields were rather low (0.002% of soluble proteins in tobacco seedlings). The authors suggested transforming tobacco plants with a plasmid that contains the code for the human γ-carboxylase and then cross the respective transgenic plants with plants transformed with the hPC gene. They also suggested putting an inducible promoter ahead of the hPC and then inducing detached tobacco leaves with the respective activator — to obtain *de novo* transgene activation. This should avoid the degradation of the hPC in the growing transgenic plants during the time lapse between the harvesting of leaves and hPC extraction. We may note here that although transgenic tobacco leaves are kosher and were repeatedly claimed to constitute a source of choice for therapeutic products — this source did not compete successfully with transgenic sows… . Velander *et al.* (1992) announced the production of hPC in the milk of transgenic swine; this latter mode of biotechnology for hPC production still awaits the tobacco challenge.

5.5.14. *β-carotene*

Those who are consuming yellow butter, made from milk of Jersey or Guernsey cows, should not worry about vitamin A deficiency. But, the people of eastern Asia, where rice is the staple food, and they consume very little cow milk (and if they do, their milk will not be from Channel-Island cows), should be concerned about vitamin A deficiency. In fact, this deficiency is causing health problems in Asia where hundreds of thousand of children suffer from blindness resulting from this deficiency. In addition, there are other less severe vision problems that result from the lack of vitamin A or its yellow-orange precursor: β-carotene. It is therefore befitting that I. Potrykus and his associates (who included investigators from China) used recombinant DNA and genetic transformation to produce β-carotene in transgenic rice (Ye *et al.*, 2000). The synthesis of β-carotene in plants requires two enzymes to create double bonds: phytoene desaturase and ζ-carotene desaturase. A third enzyme, lycopene β-cyclase (*lcy*), is also required. The first two enzymes, for the creation of double bonds, can be replaced by one bacterial enzyme, carotene desaturase.

The investigators engineered three *A. tumefaciens* binary vectors. The first vector (pB19hpc) contained, in addition to a selectable marker, genes for two enzymes: one coding for a plant phytoene synthase (*psy*) and *crtl*, a bacterial (*Erwinia*) gene coding for phytoene desaturase. A code for a transit peptide that leads the enzymes to the plastids was added in each case and promoters for expression in the endosperm were placed upstream of the two genes. This vector was intended to form lycopene in the rice endosperm. Immature rice embryos were used for *Agrobacterium*-mediated transformation and the seedlings that passed the selection (in hygromycin) were regenerated to plants. The existence of the transgenes was verified. The T_0 plants were selfed and segregated 3:1 (colored: noncolored grains). Two other vectors were constructed to serve in co-transformation with the agrobacteria. One vector, pZPsC, contained as pB19hpc the two genes *psy* and *crtl* with the same transit peptides and promoters, but lacked the selectable gene. The other vector, pZLcyH, had a selectable gene as well as *lcy*, a bacterial gene for phytoene desaturase. Plants that expressed all the three genes (*psy, crtl* and *lcy)* in their endosperm were selected. For reasons that were not clear to the investigators, the endosperms that resulted from transformation with pB19hpc contained β-carotene rather than lycopene. The other co-transformation also resulted in plants that produced grains containing β-carotene. In all the transformed grains, there were also glutenin and zeaxanthin. Although this investigation is still not finished (the resulting transgenic lines were not yet stabilized by further generations of selfed plants), the results are impressive and can lead the way to further improvements in the nutritional value of crops. The question of public acceptance of genetically modified food is of course pending but results of this kind may shift the choice of the public to acceptance.

5.6. Biolistic Transformation

Although biolistic transformation (see Chap. 2) is considered inferior, with several respects to *Agrobacterium*-mediated transformation, it was a procedure of choice to transform cereal crops in many studies. As we shall see below, a procedure to obtain a product that is already being sold commercially was based on biolistic transformation.

5.6.1. *Avidin*

Avidin is a glycoprotein found in avian, reptilian and amphibian eggs (in the egg white). This protein is composed of four identical subunits of 128 amino acids each. In chicken, there is a family of closely related avidin genes. Avidin is not a therapeutic product but we shall deal with it because it is a high-value protein that is used in diagnostic procedures in biomedicine. The common source of avidin is the egg white of chicken eggs. It is an expensive product, the gene for chicken avidin is available and because it is not a therapeutic protein, the purity requirements are not very strict. All these considerations probably combined to trigger a team of 21 (!) investigators from three commercial companies and one university to come up with a procedure to produce avidin in the grains of transgenic maize and to bring it up to the stage of commercialization (Hood *et al.*, 1997; 1999).

These investigators constructed the plasmids for transformation meticulously. They used computer-assisted procedures to synthesize the cDNA for chicken egg avidin so that this coding sequence will have the maize codon usage. Likewise, they synthesized the sequence for the barley α-amylase signal sequence to optimize it to the maize codon usage and made the appropriate nucleotide changes to enable the engineering of the plasmids. Several intermediate (commercially available) plasmids were used for optimal cloning of the maize ubiquitin 5′ region, which included a promoter, the first exon and the first intron upstream of the synthetic cDNA for avidin. A terminator sequence was cloned downstream of this cDNA. This resulted in plasmid PHP5168. In another plasmid, they engineered the *bar* gene as a selectable marker (for resistance to the herbicide bialaphos). In this plasmid (PHP610), the *bar* gene was flanked by a double 35S CaMV promoter and the *Pin*II terminator (from potato) (Fig. 23).

Cells of maize from a single immature embryo tissue were transformed by the bombardment-mediated procedure (biolistics). For that, an equimolal amount of both plasmids PHP5168 and PHP610 were mixed before bombardment and the selection was done by culture on bialaphos-containing culture medium for ten weeks. The resistant colonies were also tested for the production of avidin by ELISA and specific calli were regenerated into functional maize plants. These plants were screened on the basis of avidin production in their leaves. It should be noted that most transgenic maize plants showed male sterility and therefore were mostly propagated by cross-pollination

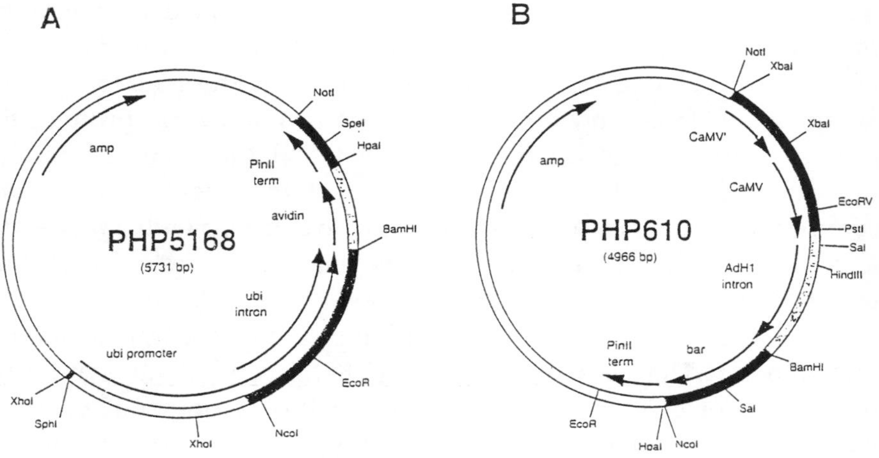

Fig. 23. Restriction endonuclease maps of plasmids used for maize transformation to generate the avidin-1 line. (A) PHP5168 is composed of the maize ubiquitin promoter and nontranslated first exon with intron, a barley α-amylase signal sequence, a gene encoding chicken avidin and the *Pin*II terminator cloned into a pBluescript SK+ plasmid backbone. (B) PHP610 is composed of the tandem CaMV promoter followed by the *Ad*H1 intron from maize, the *bar* gene and the *Pin*II terminator cloned into pBluescript SK+. amp = resistance to ampicillin (From Hood *et al.*, 1997.)

with pollen of nontransformed maize plants. It is also noteworthy that although the avidin gene and the *bar* gene were on separate plasmids, they were genetically linked in the progeny of the transgenic maize. This is a phenomenon that was reported frequently by other investigators who used biolistics for genetic transformation of cereals. The avidin was also found in the transgenic maize grains and the signal peptide was correctly cleaved from the mature protein. The level of avidin in the kernels was rather high: about 2 to 6% of the aqueous extractable protein. Tissue printing of grain tissue (with anti-avidin antibody) located the avidin in the embryo of the grain (Fig. 24) that comprises only about 20% of the grain's weight. Because separating the embryos from the endosperm is a routine procedure in maize processing, the avidin could thus be obtained in a rather concentrated form. The whole system was scaled up and the avidin production

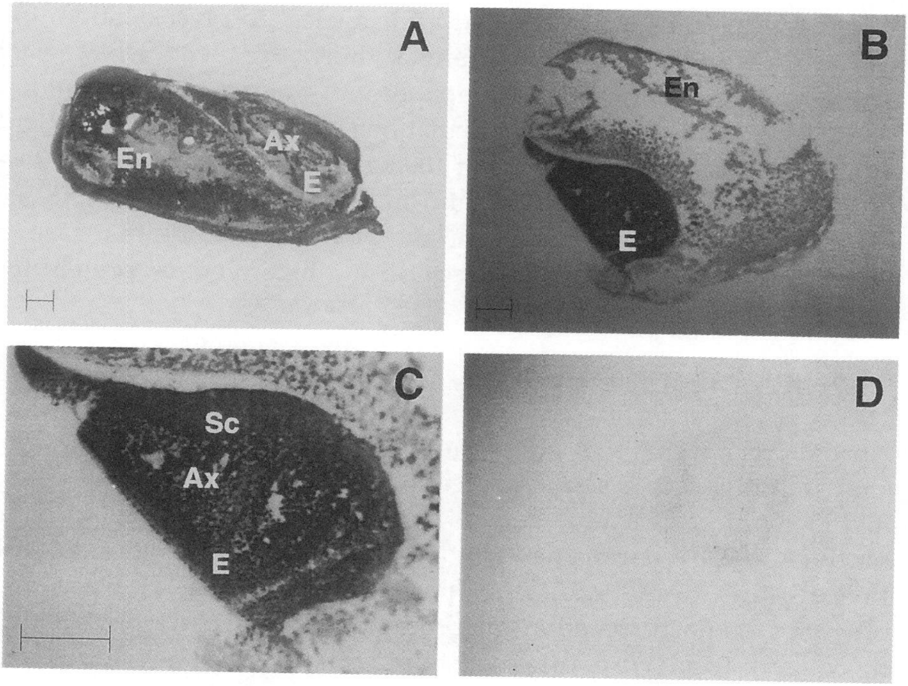

Fig. 24. Tissue prints of near-mature kernels from an avidin-positive plant [(A)–(C)] or an avidin-negative plant (D). (A) Tissue print stained for total protein with Indian ink. (B) and (C) Tissue prints incubated with the anti-avidin antibody (ICN, 1:1,000 dilution) and an alkaline-phosphatase-conjugated secondary antibody (Sigma Chemical Co., 1:1,000 dilution). Note the high concentration of avidin protein in the embryo, E. (D) Avidin-negative kernel tissue print incubated as in (B) and (C). Ax, embryonic axis; Sc, scutellum; En, endosperm. Bars represent 1 mm. (From Hood *et al.*, 1997.)

capability was moved, by appropriate breeding, into high-yielding hybrid maize. The authors then developed a processing process and found that the avidin content and functionality were not destroyed by the regular maize seed processing procedures. The mean yield of avidin in several hybrids was 230 mg/kg kernels. Amino acid sequencing showed identity between the recombinant avidin and authentic avidin.

It was revealed, as expected, that with respect to glycosylation, the two avidins differed but this difference did not reduce the biological activity and the biotin-binding capability of the recombinant avidin. These authors also went into calculations. But their calculations had a sound basis. They indicated that 100 kg of maize seed can yield 20 g of avidin and this amount is found in 900 kg of eggs (about 18,000 eggs!) The authors proudly declared: "This presents the first case in which a plant has been used to produce a heterologous protein for commercial scale".

5.6.2. *Aprotinin*

Aprotinin is a chain of 58 amino acids with a molecular mass of 6511 daltons. It is a serine proteinase inhibitor and has been used in biochemical research and as a therapeutic agent. Its current production is by purification from bovine pancreas and lung tissues. Investigators from ProdiGene Inc. (E.E. Hood, J.A. Howard and others) who participated in the production of maize-derived avidin, in collaboration with others (Zhong *et al.*, 1999) developed the production of aprotinin in maize "seeds" (again, the authors wrote "seeds" but meant kernels or grains). The strategy of engineering the plasmids for transformation, the biolistic transformation, the selection of calli, etc. were all similar to the strategy of the previous study (Hood *et al.*, 1997). Briefly, two plasmids were constructed, one with the synthetic cDNA for aprotinin, with a correct promoter and a barley α-amylase signal peptide (plasmid PHP5639). The other plasmid (PHP3528) was engineered to express the *bar* gene (for resistance to the herbicide bialaphos). Particle bombardment (biolistics) of a mixture of the two plasmids into immature embryo-derived maize cells was followed with selection of calli on bialaphos-containing culture medium and regeneration of maize plants. One putative transgenic plant with high aprotinin in its leaves was selected. Because in this transformation the transformed plants were not male sterile (as was reported for avidin-containing transgenic maize in

Hood *et al.*, 1997), the pollen of this plant was used to cross-pollinate untransformed maize plants. The progeny was self-pollinated to result in a T_2 progeny. Again, expressions of aprotinin and *bar* co-segregated, indicating linkage. Also, as in the previous study, the aprotinin was located mostly in the embryos of the transgenic maize kernels. One line was more stable in aprotinin production than other lines and the finally selected lines produced kernels with about 0.35% aprotinin in the soluble protein extract. By several criteria, the recombinant aprotinin was identical to authentic aprotinin (e.g. sequence of N-terminal amino acids, Western blot hybridization). The functional characterization (trypsin inhibition assays) of the recombinant aprotinin indicated that its activity was equal or greater than the activity of native aprotinin. The recovery efficiency of aprotinin from transgenic maize kernels was 30% or more.

5.6.3. *Stilbene synthase*

Stilbene synthase (*sts*) is a plant gene that produces stilbene-type phytoalexin, like resveratrol. Resveratrol has fungicidal potentials and was also claimed to have beneficial effects on human health (Gehm *et al.*, 1997; Jang *et al.*, 1997). Resveratrol is a component of red wine and as such the benefit of red wine on human health is advocated. While rigorous medical evidence on such effects is lacking, it probably affected positively the wine production industry. Fettig and Hess (1999) expressed the *sts* gene in transgenic wheat; by that they caused the horizontal transfer of a gene from the plant that yields the oldest beverage (remember Noah of the Jewish Bible who, after a flood of excessive water, first planted a vineyard) to a plant that is the source of our bread. Thus, we overcame our hesitation of including in this book the *sts* expression in transgenic wheat.

Fettig and Hess (1999) established a scutellar-callus culture from immature wheat embryos. After three weeks, the nonembryogenic callus was separated from the embryogenic callus and the latter was exposed to particle bombardment. A mixture of two plasmids was used for bombardment: plasmid pStil2 that contained the grapevine-derived *sts* gene followed by its own terminator and driven by a ubiquitin promoter (that also contained an exon and an intron of ubiquitin) and plasmid pAHC25 that contained the *gus* reporter gene (with the same ubiquitin promoter as pStil2 and a *Nos* terminator) as well as a cassette for the expression of the *bar* gene for selection on phosphinothricin. In another type of bombardment plasmid, pStil 2 was mixed (1:1) with the plasmid for expressing only the *gus* gene. After due selection, during culture on phosphinothricin and regeneration of resistant calli to plantlets, the plantlets were transferred to growth in the greenhouse. The rate of transformation varied between 0.02 and 0.29%. Co-transformation of the genes in the two plasmids used in each type of bombardment was high. The *sts* expression was assured for the T_0 and T_1 generations in two lines that were bombarded with pStil2 + PAHC25 (lines F and I), but the levels of resveratrol were low and could only be detected after acidic hydrolysis. Control wheat also had a low level of resveratrol but this level was still much lower than that in the transgenic wheat plants. The levels of resveratrol in the transgenic wheat lines were estimated by the authors to be about 2 μg/g of fresh weight (of leaves!) as compared to 50–100 μg/ml of wine, and no information was provided for the resveratrol level in the kernels of the transgenic wheat plants. The authors were optimistic and suggested that using a promoter for the expression of *sts* in the wheat grain would improve the level of resveratrol and they initiated a study in which they used an endosperm-specific promoter upstream of the *sts* gene. They conclude with a recommendation: "At present it is not yet possible to consume beneficial resveratrol without increasing alcohol consumption — just by enjoying red wine with resveratrol-white-bread". With the existing mood in Germany regarding genetically modified food, one has great doubts with respect to the possibility of bringing to the market bread made from transgenic wheat with elevated resveratrol content.

5.7. *Agrobacterium rhizogenes* Transformation

In this section, we shall handle exclusively the manufacture of products obtained in transgenic roots after *A. rhizogenes*-mediated transformation. This section is not merely unique with respect to the methodology of transformation but also with respect to the types of products and the plant organ from which the products are harvested.

While we previously dealt mostly with the production of peptides (short, medium or long and elaborate), such as antibodies (Chap. 3), antigens (Chap. 4) and various therapeutic proteins (in previous sections of this chapter), we shall discuss completely different types of products here. All these products can be defined as secondary metabolites of plants. Most of the products are pharmaceutics that are commonly derived from medical plants. Notably, plant-derived pharmaceutics vary vastly in their value (see Oksman-Caldentey and Hiltunen, 1996). When given at retail prices of US$ per g, caffeine, atropine, berberine and papaverine are between 0.1 to 6.5; scopolamine, cocaine, codeine and colchicine are between 17.5 and 66.0; vinblastine and vincristine are about 12,000 and 32,000, respectively (Table 2). The interest in the manufacture of pharmaceutics in transgenic hairy roots is affected by the value and market demand of each product. Most of the products can be divided into four groups according to their chemical class: quinones, phenylpropanoids, isoprenoids and alkaloids.

This section will exclusively deal with "hairy-root" cultures rather than with whole plants, as mentioned elsewhere in this book. The molecular biology of *A. rhizogenes* infection was amply described (see Chilton *et al.*, 1982; Tepfer, 1984; Durand-Tardif *et al.*, 1985) although it is still less understood than the molecular biology of *A. tumefaciens* infection. But for practical application of transforming plants, there is sufficient knowledge as what was summarized in Chap. 2 (Sec. 2.5). In short, *A. rhizogenes* also harbors a plasmid that contains a T-DNA region and this region can integrate into the nuclear genomes of plants where the genes located within this T-DNA are expressed. Once integrated, the T-DNA is transmitted to the sexual progeny. The

Table 2. The market prices of some important plant-derived compounds produced in the pharmaceutical industry.

Plant species	Product	Price[a](US$ g^{-1})
Atropa belladonna	atropine	6.45
Datura sp.	atropine	
Hyoscyamus sp.	atropine	
Duboisia sp.	scopolamine	17.50
Nicotiana sp.	nicotine	144.80
	ubiquinone-10	1,667.00
Coptis japonica	berberine	1.70
Berberis sp.	berberine	
Catharanthus roseus	ajmalicine	13.00
	vinblastine	11,900.00
	vincristine	32,350.00
Erythrozylon coca	cocaine	66.00
Physostigma venenosum	physostigmine	34.60
Pilocarpus microphyllus	pilocarpine	15.40
Papaver somniferum	sangjinarine	2,920.00
	morphine	528.00
	codeine	18.90
	papaverine	2.10
Colchicum autumnale	colchicine	36.30
Digitalis sp.	digoxin	48.30
Dioscorea sp.	diosgenin	3.10
Claviceps purpurea	ergotamine	78.15
	ergometrine	230.00
Coffea sp.	caffeine	0.08
Ephedra sinica	ephedrine	0.93

[a]Prices are based on the 1993 catalogue of Sigma chemicals.
(From Oksman-Caldentey and Hiltunen, 1996.)
Note: For comparison, the price of gold is about US$10.00 g^{-1}.

T-DNA of *A. rhizogenes* contains genes that induce "hairy roots" which are capable of growing in culture without the addition of plant growth regulators (such as auxins).

This section shall be divided into two parts; in the first part, we shall describe studies in which *A. rhizogenes* was utilized without engineering its T-DNA region. This will initiate "hairy roots" that are maintained as root cultures but no alien genes in addition to the regular T-DNA genes will be transferred to the transgenic roots. In the second part, we shall describe the relatively few studies in which the T-DNA of the *A. rhizogenes* plasmid was first engineered to contain and express additional transgenes. The few cases in which only a reporter gene was added to the *A. rhizogenes* plasmid will be included in the first part of this section. Overviews on the production of pharmaceutics in transgenic root cultures were provided by Oksman-Caldenty and Hiltunen (1996), Flores *et al.* (1997), Oksman-Caldentey and Arroo (1999), and Shanks and Morgan (1999). A list of genera and

Table 3. Plant genera and their family affiliations that are used in hairy-root cultures for the production of pharmaceutics.

Genus	Family
Artemisia	(Compositae)
Atropa	(Solanaceae)
Calystegia	(Solanaceae)
Catharanthus	(Apocynaceae)
Cinchona	(Rubiaceae)
Coleus	(Labiatae)
Datura	(Solanaceae)
Digitalis	(Scrophulariaceae)
Duboisia	(Solanaceae)
Echinacea	(Asteraceae)
Glycyrrhiza	(Papilionaceae)
Hyoscyamus	(Solanaceae)
Lawsonia	(Lythraceae)
Lippia	(Verbenaceae)
Lithospermum	(Boraginaceae)
Nicotiana	(Solanaceae)
Panax	(Araliaceae)
Paulownia	(Bignoniaceae)
Peganum	(Zygophyllaceae)
Rubia	(Rubiaceae)
Salvia	(Lamiaceae)
Scopolia	(Solanaceae)
Scutellaria	(Lamiaceae)
Tagetes	(Compositae)
Valeriana	(Valerianaceae)

Table 4. The secondary metabolite production of the hairy root cultures of some important medicinal plants.

Plant	Product	Content[a]
Atropa belladonna	atropine	0.37%
	atropine + hyoscyamine	0.95%
	scopolamine	0.3%
Catharanthus roseus	ajmalicine	4 mg/g
	catharanthine	2 mg/g
	serpentine	2 mg/g
	vindoline	0.4 mg/g
	vinblastine	0.05 μg/g
Cinchona ledgeriana	quinine	25 μg/g f.w.
Datura candida	scopolamine	0.57%
Datura innoxia	hyoscyamine	1.7 mg/g
Datura stramonium	hyoscyamine	0.3%
	scopolamine	0.56%
Duboisia hybrid	scopolamine	2.5 mg/g
	hyoscyamine	2.1 mg/g
Duboisia leichhardtii	scopolamine	1.8%
	scopolamine	90 mg/1
Duboisia myoporoides	scopolamine	0.24%
Echinacea purpurea	alkamides	0.03%
Glycyrrhiza uralensis	glycyrrhizin	4.7%
Hyoscyamus albus	hyoscyamine	8 mg/g
	scopolamine	4.6 mg/g
Hyoscyamus niger	hyoscyamine	12.5 mg/g
Hyoscyamus muticus	hyoscyamine	9 mg/g
Lippia dulcis	hernandulcin	1.5 mg/g
Lithospermum erythrorhizon	shikonin	68 mg/g
Nicotiana rustica	nicotine	0.65 mg/g f.w.
	nicotine	4 μmol/g f.w.
Nicotiana tabacum	cadaverine	0.7 mg/g
	anabasine	3.4 mg/g
Panax ginseng	ginsenosides	0.95%
Peganum harmala	β-carbolines	17 mg/g
Salvia miltiorrhiza	tanshinones	19 mg/g
Scopolia japonica	scopolamine	0.5%
	hyoscyamine	1.3%
Scopolia tangutica	hyoscyamine	0.52 mg/g
Tagetes patula	thiophenes	9 mg/g

[a]Contents of secondary product presented on dry weight basis if not otherwise indicated; f.w., fresh weight. (From Oksman-Caldentey and Hiltunen, 1996.)

their family affiliations that are used in hairy-root cultures for the production of pharmaceutics is given in Table 3. The contents of secondary metabolites in the hairy roots of several medicinal plants are provided in Table 4. An update (for 1998) on metabolites produced in hairy-root cultures is provided in Table 5. Although the molecular aspects of *A. rhizogenes* infection have just started to be understood only about 15 to 20 years ago, we should recall that the phenomenon of hairy-root formation after infection with *A. rhizogenes* was already published about 70 years ago (Ricker *et al.*, 1930). But it was only after information on the insertion of the T-DNA of *A. rhizogenes* into the host chromosome became available that studies were initiated to use

Table 5. Examples of metabolites in hairy-root cultures.

Genus	Metabolite	Function	References*
Artemisia	artemisinin	anti-malarial	[32, 38]
Atropa	tropane alkanoids	anticholinergic	[45]
Beta	betalains	coloring agent	[33]
Brugmansia	tropane alkanoids	anticholinergic	[42]
Catharanthus	indole alkanoids	antihypersensitive	[19–21,23,24,40]
Coleus	soreskolin	antihypersensitive	[46]
Datura	tropane alkanoids	anticholinergic	[47]
Glycyrrhiza	isoprenylated flavanoids, polysaccharides	antimicrobial, antioxidant, immunomodulatory	[6,48]
Henna	lawsone	coloring agent	[8]
Hyoscamus	tropane alkanoids	anticholinergic	[39,41]
Lithospermum	benzoquinones	coloring agents	[7]
Panax	polyacetylene analogs	unknown	[5]
Paulownia	verbascoside	antibacterial, antiviral	[49]
Pimpinella	essential oils	flavorings, perfumes	[50]
Scutellaria	flavonoids	antibacterial, antioxidant	[14]
Solanum	solasodine	steroidal drug, precursor	[9]
Trachelium	polyacetylenes	unknown	[51]
Trichosanthes	ribosome inactivating protein	antiviral, antifungal	[10]
Valeriana	valepotriates	sedative and tranquilizing activities	[52]

*From Shanks and Morgan (1999); see references in this publication.

hairy roots for the production of secondary metabolites. As we shall see below, for about six years (from 1986 to 1992), investigators used the T-DNA of *A. rhizogenes* without engineering it further. It was only thereafter that there were publications of studies in which this T-DNA was first engineered to render the production of secondary metabolites in hairy roots more efficient, and *A. rhizogenes* with engineered plasmids was used to infect medicinal plants.

5.7.1. *Induction of hairy root cultures in medicinal plants with unmanipulated A. rhizogenes*

5.7.1.1. *The first attempts*

The term medicinal plants encompasses a vast number of plant species. We shall focus on specific species in which genetic transformation was performed by *A. rhizogenes* in order to obtain root cultures. An overview of such studies performed between 1986 and 2000 is provided in Table 6, where the publications are listed in chronological order.

The first reviewed publication of a study to induce hairy roots for the production of secondary metabolites was submitted as early as December 1985 by a group of investigators from Norwich, England (Hamill *et al.*, 1986). Norwich is a source of several pioneering plant investigations, some of these were noted by us previously. Is this a counterflow of culture and scholarship to what happened 12 centuries earlier when repeated barbaric invasions were westbound from Yarmouth to Norwich? These investigators germinated seeds of *Nicotiana rustica* and of a beet cultivar (*Beta vulgaris*), and then infected the young plants with a strain of *A. rhizogenes*. The roots that were induced at the sites of inoculation were transferred to liquid culture medium that contained an antibiotic (ampicillin) but no growth hormones. The growth of roots was maintained in 250 ml flasks (shake culture) with transfer to fresh medium at two weeks intervals.

The integration of the T-DNA from *A. rhizogenes* was verified in the roots by Southern blot hybridization. The weight of the cultured

Table 6. Examples of the establishment of hairy root cultures for the production of secondary metabolites of pharmaceutical interest.

Species	Publication
Beta vulgaris, Nicotiana rustica	Hamill *et al.*, 1986
Atropa belladonna	Kamada *et al.*, 1986
Atropa belladonna, Calystegia sepium	Jung and Tepfer, 1987
Dubosia myoporoides	Deno *et al.*, 1987
Nicotiana rustica	Furze *et al.*, 1987
Panax ginseng	Yoshikawa and Furuya, 1987
Datura stramonium	Payne *et al.*, 1987
Catharanthus roseus	Endo *et al.*, 1987
Catharanthus roseus	Parr, 1988
Datura stramonium, Hyoscyamus niger	Jaziri *et al.*, 1988
Cinchona ledgeriana	Hamill *et al.*, 1989
Datura candida	Christen *et al.*, 1989
Duboisia leichhardtii	Mano *et al.*, 1989
Glycyrrhiza uralensis	Saito *et al.*, 1990a
Digitalis purpurea	Saito *et al.*, 1990b
Tagetes patula	Mukundan and Hjortso, 1991
Hyoscyamus muticus	Oksman-Caldentey *et al.*, 1991
Lippia dulcis	Sauerwine *et al.*, 1991
Hyoscyamus albus, Scopolia tangutica, Datura innoxia	Shimomura *et al.*, 1991a
Lithospermum erythorhizon	Shimomura *et al.*, 1991b
Echinacea purpurea	Trypsteen *et al.*, 1991
Peganum harmala	Berlin *et al.*, 1992
Duboisia leichhardtii	Muranaka *et al.*, 1992
Valeriana officinalis	Gränicher *et al.*, 1992
Catharanthus roseus	Bhadra *et al.*, 1993
Salvia miltiorrhiza	Hu and Alfermann, 1993
Hyoscyamus albus	Sauerwein and Wink, 1993
Solanum aviculare	Subroto and Doran, 1994
Duboisia myoporoides	Yukimune *et al.*, 1994
Rubia peregrina	Lodhi *et al.*, 1996
Lawsonia inermis	Bakkali *et al.*, 1997
Atropa belladonna	Aoki *et al.*, 1997
Scutellaria baicalensis	Nishikawa and Ishimaru, 1997
Paulownia tomentosa	Wysokinska and Rozga, 1998
Coleus forskohliii	Sasaki *et al.*, 1998
Valeriana wallichii	Banerjee *et al.*, 1998
Solanum aviculare	Kittipongpatana *et al.*, 1998
Artemisia annua	Liu *et al.*, 1998
Catharanthus roseus	Morgan and Shanks, 2000

Note: The *A. rhizogenes* plasmid did not contain a transgene that specifically affected the metabolic pathway.

roots increased during the first ten days of culture, then the weight remained constant. For harvesting secondary metabolites, the cultures were maintained for 20 days. In the beet-root cultures, the content of the red pigments, betaxanthin and betaganin, rose to peaks of 0.7 and 1.3 mg per g fresh weight, respectively. Thereafter, there was lysis of the roots and the pigments were released into the culture medium. A similar trend was observed in hairy-root cultures of *N. rustica.* There was a slow increase in root weight up to the 9th day of culture and then there was faster growth until the 17th day. The nicotine content per g fresh weight hardly increased till after the 9th day and the content peaked at the 17th day (0.3 μg per g fresh weight). The secondary metabolites in these root cultures were at similar levels as in the respective intact plants.

While beets and *N. rustica* are not considered medicinal plants, a team of investigators from Japan (Kamada *et al.*, 1986) investigated the alkaloid production in the hairy roots of a real medicinal plant: *Atropa belladonna.* It should be noted that there has been a long tradition in Japan of producing pharmaceutics in plant cell cultures. Therefore, the shift from cell culture to the culture of hairy roots was a logical transition. These investigators raised axenic *A. belladonna* plants and then inoculated the stems (by a needle) with a suspension of *A. rhizogenes.* After a few weeks, roots emerged at the sites of inoculation. These roots were first transferred to solidified (agar) medium without growth hormones, but with carbenicillin to kill the agrobacteria. Thereafter, root sections were transferred to liquid medium (100 ml flasks, shake cultures) that also lacked growth hormones. The roots were harvested after four weeks in shake culture. The levels of atropine and scopolamine reached 0.37 and 0.024%, respectively, in the cultured hairy roots. For comparison, the levels of atropine and scopolamine in field-grown *A. belladonna* were found to be 0.34 and 0.008%, respectively, and almost no alkaloids were found in roots that were cultured without transformation.

A similar study was conducted by Jung and Tepfer (1987). These investigators established hairy-root cultures of *A. belladonna* and *Calystegia sepium* and followed the increase of root mass and tropane alkaloids. The hairy roots were induced by a similar method as noted above but the contribution of the latter investigators was mainly the improvement of the culture technolology. The investigators compared batch culture in flasks (300 ml flasks with 100 ml culture medium) to fermenters of 2 or 30 liters (with 1 or 20 liters of medium). They found that culture in 2-liter fermenters was very efficient: the biomass of transformed *C. sepium* roots in 2-liter fermenters reached 0.8 g/l/day (dry weight). And the yield of alkaloids was 2.3 mg/l/day. It should be noted that untransformed *C. sepium* roots did not grow at all in fermenters and reached only 0.02 g/l/day (dry weight) in flasks. Less growth was observed in fermenter-cultured hairy roots of *A. belladonna*; nevertheless, the alkaloid content in these roots exceeded the alkaloid content in untransformed roots by several folds.

Duboisia myoporoides is a natural source for scopolamine and hyoscyamine. Thus, Deno *et al.* (1987) used this species to induce hairy roots. These investigators had previous experience with untransformed roots of *D. myoporoides*. Such roots could be grown in liquid cultures provided they were supplemented with indolebutyric acid. The aim of their investigation was to find out whether hairy-root cultures are a better source of scopolamine than cultures of untransformed roots. The establishment of hairy-root culture was similar to that of previously mentioned studies. Only batch cultures were made. The hairy roots showed good growth in these cultures. The total content of both scopolamine and hyoscyamine in the hairy roots was similar to that found in untransformed root cultures (about 1% of dry weight). But there was a shift in ratio between scopolamine and hyoscyamine: in the hairy roots, there was a reduction in scopolamine and an increase in hyoscyamine. Such a change was obviously not desirable. The addition of several growth regulators within the physiological range did not increase the content of tropane alkaloids in the hairy roots.

5.7.1.2. *Is it possible to select for high yield of secondary metabolites in hairy roots?*

There is no proven answer to this question but we tend to claim that it is. First, evidence comes from the unpublished experience of one of us (Esra Galun and Dvora Aviv, unpublished). In the late seventies, we were asked to breed a line of *Datura (Brugmansia) sanguinea* that will be rich in scopolamine. Seeds were collected in Latin American countries and given to us. We were surprised to find that even within a small population of plants that germinated from these seeds, the variability of scopolamine content was vast: between 0.1 and 2.0% of the dry weight of leaves. Crosses between individual high yielding plants quickly resulted in high-yielding lines. In retrospect, this should not surprise us because no previous selection was performed for high scopolamine content in *Datura sanguinea*. Moreover, *D. sanguinea* has (gametophytic) incompatability; thus self-pollination does not take place and heterogenicity is maintained in the wild population. Another indirect evidence comes from sugarbeet breeding. Before breeding started (some 150 years ago), the sucrose content of sugarbeet was only a few percent (of fresh weight). Presently, due to a lot of breeding, the content of sucrose was raised to 18%! The irony is that in spite of this several-fold increase in sugar content, the production of sugar from sugarcane is much cheaper than from sugarbeet. The take-home lesson is that if one is dealing with the establishment of hairy roots from a medicinal plant, there should first be a phase of selection so that the induction of such roots be performed with individual plants that excel in high content of the secondary metabolite.

The above note is an introduction to a study by Furze *et al.* (1987). They assumed that conversion of plant cells to protoplasts and then regeneration of hairy roots from these protoplasts will cause somaclonal variability among the individual lines of the hairy root cultures so that some of them may reach a high secondary metabolite level. These investigators took hairy roots of *Nicotiana rustica* and obtained calli from their protoplasts (by appropriate enzyme treatment). The calli were further cultured until they again produced hairy roots. The latter were cultured as individual lines and their nicotine content was evaluated. Indeed there was a variability among the protoplast-derived root clones with respect to the nicotine content. Furthermore, these clones were also variable with respect to root growth in culture.

A selection of hairy-root clones was also performed by Mano *et al.* (1989), with *Duboisia leichhardtii.* These investigators obtained 700 root cultures of which 45 were grown in medium without growth hormones. They then further selected for growth rate and scopolamine production. One clone of hairy roots produced 1.8% scopolamine (on dry weight basis).

It should be noted that *Nicotiana* is one of the relatively few genera in which cells can be easily converted to protoplasts and then the protoplasts can divide, produce calli and ultimately functional plants. Thus, selection for high content of secondary metabolites based on individual protoplast-derived root-hair lines will be possible in only very few medicinal plant genera. Moreover, unless one is dealing with a homozygous cultivar, it makes more sense to start the selection at a previous stage: first to test numerous plants and select those that are the highest producers among a large population of individual plants. Possibly, some breeding among the selected plants is worthwhile.

Bhadra *et al.* (1993) claimed that at least with respect to two different medicinal plants, having a higher indole alkaloid content in the intact plant does not prove that the high-content species will also be a better alkaloid producer in hairy-root cultures. They tested four different cultivars of *Catharanthus roseus* and produced from them 150 lines of hairy roots. They finally arrived at two hairy-root lines that contained vindoline levels that were *three orders of magnitude* greater than the minute levels of vindoline reported in cell cultures of *C. roseus.* It should be recalled that vindoline and another compound, derived from *C. roseus,* are required to synthesize the expensive anticancer drug vinblastine.

Yukimune *et al.* (1994) studied the effect of selection among hairy- root lines of *Duboisia myoporoides* to increase the scopolamine content. They noted that the scopolamine content of untransformed root cultures in this species is about 1% (of dry weight). Several rounds of selection did not change the level of scopolamine in untransformed root cultures. On the other hand, selection of hairy-root cultures did cause an increase of scopolamine level and up to 3.2% scopolamine was detected in one line. But, repeated selection for scopolamine also reduced the yield of roots from 7 to 2 g/l liquid medium. These authors concluded that because some changes (e.g. addition of growth hormones) will increase the growth rate of roots but reduce the scopolamine level, a two-stage culture could be optimal: first to grow the hairy roots in conditions that are optimal for growth and then change the medium to cause high level of scopolamine production. Results of such experiments are still pending. Possibly, there was a continuation in the pharmaceutical industry without information to the public.

A different approach was followed by Aoki *et al.* (1997) to select for high alkaloid levels in *Atropa belladonna.* These investigators first induced hairy roots and obtained 35 lines. Of these, five were selected for further study. The five lines showed similar levels of total alkaloids but differed considerably with respect to the ratio between scopolamine and hyoscyamine. Each of these five hairy-root lines was regenerated back into plants and then the alkaloid content in the roots of the latter plants was analyzed. Again, there was no dramatic difference in the total alkaloid content of the roots of plants that were derived from the five different hairy-root lines. But, the ratio of scopolamine to hyoscyamine differed: in the roots of one regenerated line, there was almost no scopolamine but 0.05% (dry weight) of hyoscyamine while in another line, the ratio of scopolamine to hyoscyamine was about 3:1.

Hyoscyamus is another genus where protoplasts can be isolated from tissues or cells and then cultured, and ultimately produce functional plants. Sevon *et al.* (1998) first established 100 lines of hairy-root cultures from *H. muticus* and found great variation among these lines in hyoscyamine content (between 0.03 and 0.6% of dry weight). In all cases, the ratio of hyoscyamine to the more valuable scopolamine was very high. Then, the best hairy-root line was used to isolate protoplasts and from these, 171 hairy-root lines were redifferentiated. The variation in hyoscyamine content was again very great among these lines (0.04 to 1.5% of dry weight). Once a hairy-root line was established, its hyoscyamine production remained stable although, the levels of scopolamine remained very low in all the protoplast-derived hairy-root lines (the highest yielding line gave 0.13 mg/g dry weight).

5.7.1.3. *Production of high-value therapeutics*

Up to now, we were dealing with relatively inexpensive secondary metabolites. The Norwich team noted above went on with their studies but focused on the more expensive therapeutics: quinine (from *Cinchona ledgeriana)* and vinblastine (and other related alkaloids, from *Catharanathus roseus*). These studies indicated (Parr *et al.*, 1988; Hamill *et al.*, 1989) that it is possible to obtain hairy-root cultures from *C. ledgeriana* and *C. roseus* but the yield of therapeutic alkaloids in these cultures was rather low.

There is an interesting intercontinental issue concerning the plant sources for the production of steroidal hormones and antimalarial drugs. Three continents "compete" with their respective endogenous plant species. Latin America furnished *Cinchona ledgeriana,* the source of quinine, while China furnished *Artemisia annua,* the source of artemisinin, another antimalarial compound (used as such or after modifications.). As for steroidal hormones, Latin America came up with *Dioscorea* sp. while another source of these steroids is the Chinese species *Solanum avicuale,* that produces salsodine — a compound that is used to produce the same steroidal hormones.

As noted above, an alternative source for an antimalarial drug is *Artemisia annua*, a Chinese medical plant. Liu *et al.* (1998) from Beijing, China, established root-hair cultures of *A. annua* to render artemisinin production commercially feasible. They grew the hairy roots in flasks or in fermenters. Three types of fermenters were used: (a) bubble column; (b) modified bubble column; and (c) modified inner-loop airlift bioreactor (Fig. 25). The bioreactors contained 2 liters of liquid culture medium and were maintained in 12 h light per day. The growth (expressed as dry weight of roots) and artemisinin production among the three bioreactors were best in the modified inner-loop airlift bioreactor, providing 27 g/l dry weight of roots (after 20 days of culture) and 536 mg/l of artemisinin. This is already an impressive achievement. Further improvement in the bioreactors, in the medium composition and possibly in the source of the hairy roots could bring the system to commercial feasibility.

Solanum avicuale, as noted above, is a source of steroidal alkaloids such as solasodine. The latter is a precursor for the production of steroidal hormones (e.g. progesterone, cortisone). This plant is a native species in Australia (and New Zealand). Subroto and Doran (1994) from Sydney, Australia, studied the production of steroidal alkaloids in hairy roots of *S. avicuale.* In regular soil-grown *S. avicuale* plants, the steroidal alkaloids are found mainly in the leaves and fruits. These authors found that hairy roots can produce such alkaloids and the yields in shake cultures and in air-driven bioreactors reached about 30 mg/g (dry weight) of *solasodine equivalents.* This is very near the solasodine equivalents in leaves and fruits of regular plants. The authors assumed that the limitation for even higher yields is the lack of oxygen in the culture medium.

A further study in the same system was performed by Kittipongpatana *et al.* (1998) from Philadelphia, USA. These investigators also established hairy-root cultures of *S. aviculare* but focused on real solasodine (rather than solasodine equivalents). Several modifications of the liquid culture composition were tested but there was no dramatic increase in solasodine resulting from any of these modifications [however, comparison is difficult because Subrato and Doran (1994) measured "*solasodine equivalents*"].

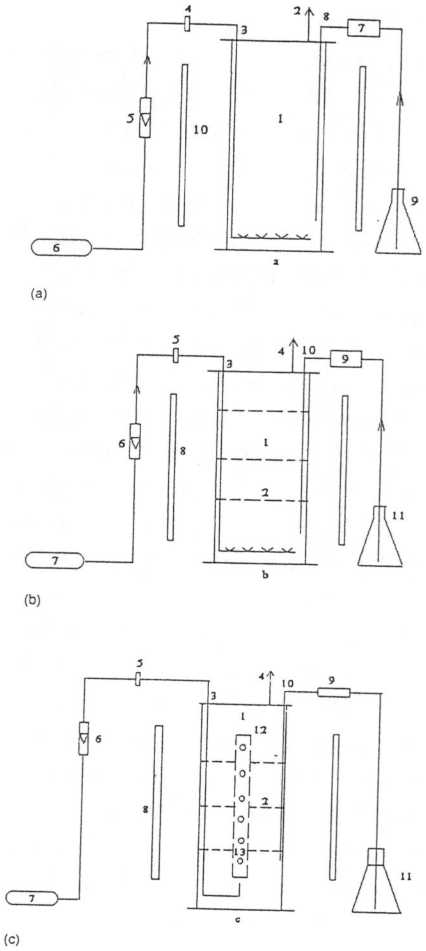

Fig. 25. (a) Diagram of the bubble column: 1. bioreactor, 2. air outlet, 3. air inlet, 4. air filter, 5. flowmeter, 6. air storage tank, 7. peristaltic pump, 8. sterile water inlet, 9. sterile water bath, 10. 40 W fluorescent lamp. (b) Diagram of the modified bubble column: 1. bioreactor. 2. stainless mesh, 3. air inlet, 4. air outlet, 5. air filter, 6. flowmeter, 7. air storage tank, 8. 40 W fluorescent lamp, 9. peristaltic pump, 10. sterile water entrance, 11. sterile water bath. (c) Diagram of the modified inner-loop airlift bioreactor: 1. bioreactor, 2. stainless mesh, 3. air inlet, 4. air outlet, 5. air filter, 6. flowmeter, 7. air storage tank, 8. 40 W fluorescent lamp, 9. peristaltic pump, 10. sterile water inlet, 11. sterile water bath, 12. inner loop, 13. holes. (From Liu *et al.*, 1998.)

5.7.1.4. *Sesquiterpenes and light*

Sauerwein *et al.* (1991) looked for a sweet compound in hairy roots of *Lippia dulcis.* These investigators looked at the production of the sesquiterpene hernandulcin and other mono- and sesquiterpenes in hairy roots. For that, they established hairy-root cultures of *L. dulcis* and grew them in liquid shake cultures. When cultured in the dark, no mono- or sesquiterpenes were produced in the roots but when cultured at a regime of 16 h light per day, these secondary metabolites were detected. Moreover, the production of hernandulcin started to increase suddenly after ca 24 days of culture reaching about 250 μg/g dry weight. Further increase of hernandulcin was observed when chitosan and naphthaleneacetic acid were added to the medium, resulting in a level that exceeded 1 mg/g dry weight. Interestingly, Saito *et al.* (1990b) who grew hairy roots of *Digitalis purpurea* found that cardenolides (e.g. digitoxin) production in these roots was correlated with greening of the roots. They used a 16 h light per day regime for hairy-root culture.

Culture in light should not be regarded as generally favoring secondary metabolite production in hairy roots. Evidence that a secondary metabolite is produced in dark-cultured but not in light-cultured hairy roots comes from an investigation that aimed to produce *henna.* Dyeing properties as well as the antispasmodic and bacteriostatic properties are attributed to lawsone, a 2-hydroxy-1,4-naphthoquinone. Henna is the powder derived from leaves of a shrub, *Lawsonia inermis,* that is cultivated in North Africa, the Middle East, Yemen and India. The powder is converted to paste and used to stain hands and hair, especially of brides before marriage. Bakkali *et al.* (1997) found that the hairy roots of *L. inermis* will produce lawsone in the dark but not in the light.

5.7.1.5. *Verbascoside from hairy roots rather than from an ornamental tree*

The ornamental "Empress Tree" (*Paulownia tomentosa*) is a native of China, Korea and Japan. In China, the tree is used in folk medicine and in an almost unlimited number of treatments, such as for gonorrhea, bruises and erysipelas. When detailed pharmacological tests were performed, it was revealed that *P. tomentosa* contains phenylpropanoid glycosides such as verbascoside in its leaves. Verbascoside was suggested to have antibacterial, antiviral and antiproliferative (in mammalian cells) activities. This compound may also have analgesic and antihypertensive properties and possibly be a potential DOPA agonist. In brief, *P. tomentosa* was claimed to contain a wide range of pharmacological properties and does contain verbascoside. The latter by itself could be useful in medicine. With all these attributes, it is no wonder that Wysokinska and Rozga (1998) went on to investigate what hairy roots of *P. tomentosa* can produce. When they established cultures of hairy roots and tried several media, they found that the production of verbascoside in these hairy roots can reach a very high level: 9.5% of dry weight. This level is double the amount found in the roots of 4.5 months old plants.

5.7.1.6. *Shikonin — A red and therapeutic naphthoquinone*

Shikonin is a red compound with therapeutic attributes (antibacterial, stimulation of granulation tissue) and cosmetic application, and it is rather popular in Japan. Indeed, it was the first secondary metabolite of pharmaceutical value that was commercially produced in cell cultures. The goal of Shimomura *et al.* (1991b) was thus to investigate if shikonin can be produced efficiently in hairy roots. These investigators therefore established hairy-root cultures from *Lithospermum erythrorhizon*. The cultured hairy roots indeed produced shikonin even in solidified (agar) medium (Fig. 26). The hairy roots were transferred to long-term culture in an airlift fermenter. The fermenter was fed with fresh medium periodically and the liquid medium was passed periodically through a bypass to a column of Amberlite XAD-4 that absorbed the shikonin. By this procedure, an almost constant yield of 5 mg/day of shikonin could be harvested from a 2-liter airlift fermenter over a 220-day period. In this specific case, the secondary metabolite was released to the medium and the removal of shikonin enabled further release from the hairy roots. The beauty of the system is that you really see

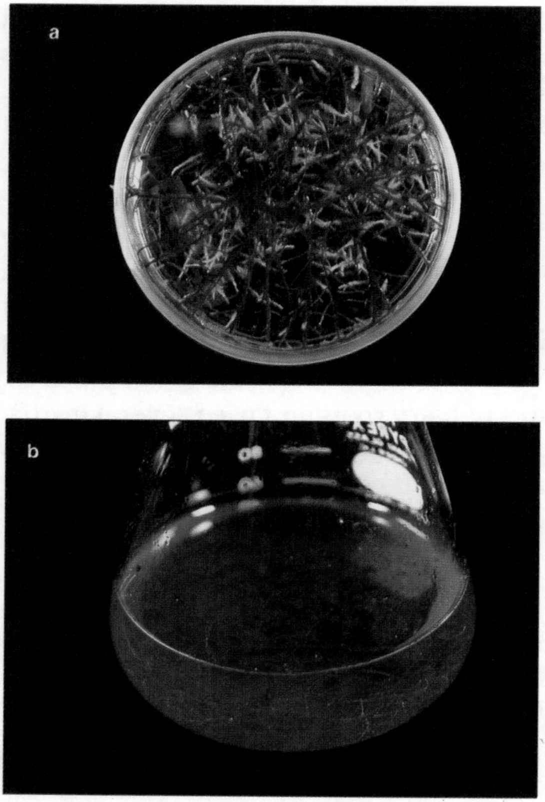

Fig. 26. (a) *Lithospermum erythrorhizon* hairy roots cultured on solid medium at 25°C in the dark for four weeks. (b) *L. erythrorhizon* hairy roots cultured in liquid medium at 25°C in the dark for four weeks. (Courtesy of Dr. Koichiro Shimomura, Tsukuba, Japan.)

Table 7. Examples of the establishment of hairy-root cultures for the production of secondary metabolites of pharmaceutical interest.

Genus	Enzyme encoded in additional gene	Product	Authors
Atropa	hyoscyamine-6β-hydroxylase	scopolamine	Yun *et al.*, 1992; Hashimoto *et al.*, 1993
Nicotiana	lysine decarboxylase	cadaverine	Fecker *et al.*, 1993
Lithospermum	chorismate pyruvate lyase	shikonin	Sommer *et al.*, 1999
Hyoscyamus	hyoscyamine-6β-hydroxylase	scopolamine	Jouhikainen *et al.*, 1999

Note: Enhancement of product was achieved by genetic engineering of the *A. rhizogenes* plasmid.

what happens due to the strong pigmentation of shikonin. It should be noted that only very few investigators can keep a plant tissue fermenter axenic for 220 days. It requires high professional skill!

5.7.2. *Induction of hairy root cultures in medicinal plants with genetically manipulated A. rhizogenes*

A summary of these studies is provided in Table 7.

5.7.2.1. *Enhancement of scopolamine production in the hairy roots of Atropa belladonna*

The tropane alkoloids, hyoscyamine (and its racemic form, atropine) and scopolamine, are important therapeutic compounds; they are anticholinergic agents that act on the parasympathetic nerve system. Atropine also serves in the daily treatment of tachyarrhythmia. Because these compounds differ in their actions on the central nervous system, scopolamine is much more in demand (and higher in value) than hyoscyamine. Scopolamine is also used as scopolamine-N-butyl bromide in the treatment of gastric disorders and serves as an antidote against nerve gas. But, in most plants that produce tropane alkaloids, the level of hyoscyamine is much higher than the level of scopolamine. Scopolamine is formed in plants from hyoscyamine in a pathway in which the enzyme hyoscyamine-6β-hydroxylase (H6H) is active (Fig. 27). Actually, the H6H is active in two steps. These considerations prompted Yasayuki Yamada (then at Kyoto University) and his associates (Hashimoto *et al.*, 1993) to attempt the enhancement of scopolamine production in hairy roots of *Atropa belladonna* by the expression of the gene encoding H6H. The gene for H6H was isolated previously in Yamada's laboratory so that it could be flanked by a 35S CaMV promoter and a *Nos* terminator and inserted into the T-DNA of a binary plasmid (pHY8) of *A. rhizogenes.* pHY8 also contained a cassette for resistance to kanamycin (*npt*II) as selectable marker. The engineered plasmid was transferred into an *A. rhizogenes* strain and the latter was then used to infect *A. belladonna.* Hairy-root cultures were obtained, of which seven were resistant to kanamycin. The

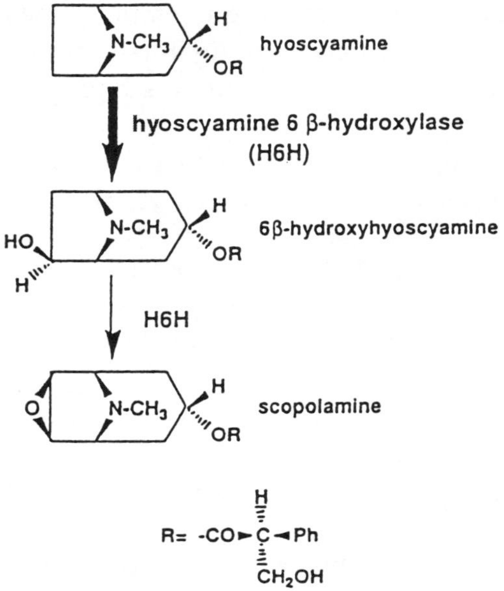

Fig. 27. Biosynthetic pathway from hyoscyamine to scopolamine. Scopolamine is formed from hyoscyamine via 6β-hydroxyhyoscyamine. H6H catalyzes the hydroxylation of hyoscyamine to 6β-hydroxyhyoscyamine, as well as the epoxidation of 6β-hydroxyhyoscyamine to scopolamine. (From Yun *et al.*, 1992.)

insertion of the transgene into the cultured hairy roots was verified and in one line (T2) of the transformed hairy roots, the level of H6H was fivefold higher than in "wildtype" hairy roots (transformed with *A. rhizogenes* that did not contain the *npt*II–H6H chimeric transgene), and the activity of H6H in T2 was likewise enhanced. The investigators determined the content of hyoscyamine, 6β-hydroxyhyoscyamine and scopolamine in T2 hairy-root cultures as well as in three "wildtype" hairy roots. The major alkaloid in the latter hairy roots was hyoscyamine (several folds higher than either scopolamine or 6β-hydroxyhyoscyamine) while in T2, the level of scopolamine was about the same as the other two alkaloids. But, even in T2, the absolute level of scopolamine was rather low, about 0.15% of dry weight. The authors also analyzed the alkaloid content of another line of hairy roots: H1R. This latter line did produce about twice the amount of scopolamine, reaching 0.3% of dry weight. The H1R line was obtained by a different strategy. *A. belladonna* plants were first transformed with a disarmed *A. tumefaciens* that contained pHY8. A transgenic *A. belladonna* plant was thus obtained. This plant was reinfected with a "wildtype" *A. rhizogenes* and, respectively, a hairy-root line was obtained. This H1R line also contained the H6H gene and elevated H6H activity. It should be noted that in a parallel study of this laboratory (Yun *et al.*, 1992), *A. belladonna* leaf discs were transformed with an *A. tumefaciens* culture that harbored pHY8. Among the transgenic plants, one expressed the H6H and contained a high ratio of scopolamine to hyoscyamine. This plant was self-pollinated. In the progeny plants, the alkaloids in the leaves and stems were almost exclusively scopolamine. The level of scopolamine in one of the progeny plants reached 1.2% of dry weight even though the scopolamine level in the roots was only about 0.1% of dry weight. Hence, these studies indicate that it is possible to interfere in a metabolic pathway that produces secondary metabolite by the expression of a transgene that encodes an enzyme of the pathway.

A few later studies followed the publications of Yamada and associates. One of these concerned a secondary metabolite which is neither a medicine nor a health product; actually a compound of ill-scent and *post-mortem* existence... cadaverine. A German team (Fecker *et al.*, 1993) added the bacterial gene coding for lysine decarboxylase (*ldc*) to a binary vector and transformed *N. tabacum* plants. Hairy-root cultures were obtained and the "best" lines of the tobacco hairy roots contained 700 μg cadaverine/g dry weight, while normally, roots contain only 50 μg/g dry weight. Feeding of lysine to the root cultures further increased the level of cadaverine.

In a study with a similar goal as that of the Yamada team, investigators from London and Leiden (Lodhi *et al.*, 1996) collaborated to integrate a bacterial gene that encodes isochorismate synthase into an *A. rhizogenes* vector. Plants of *Rubia peregrina* were transformed with the engineered *A. rhizogenes* and hairy-root cultures were obtained. The hairy roots that expressed the bacterial enzyme had elevated levels of anthraquinones.

The last two examples of interference with the metabolic pathway for secondary metabolites again deal with real pharmaceutical products: shikonin and scopolamine.

Sommer *et al.* (1999), in a binational study (Germany and Japan), focused on a bacterial gene (*cpl*) encoding chorismate pyruvate lyase. This enzyme converts chorismate into 4-hydroxybenzoate (4HB) (Fig. 28). The rationale of these investigators was that adding *cpl* will shortcut the synthesis of shikonin. They therefore introduced the *cpl* gene into the hairy root of *Lithospermum erythrorhizon.* To express the transgenic enzyme in chloroplasts, the investigators engineered a coding sequence for a chloroplastic transit peptide upstream of the *cpl* sequence. The transformed hairy roots "collaborated" only halfway: they did produce more 4HB; it was estimated that the increase of 4HB was probably 20% but there was no significant increase in the final product, shikonin.

Our final example comes from the north: Finland (Jouhikainen *et al.,* 1999). These Finnish investigators were dealing with the same metabolic pathway as Hashimoto *et al.* (1993), namely the production of scopolamine from hyoscyamine. But, the Finnish team chose *Hyoscyamus muticus* (from Egypt) rather than *A. belladonna* for the production of hairy roots. Incidentally, *H. muticus* has a close relative which is a native of the Sinai desert: *H. boveanus.* The latter has probably an even higher alkaloid content. It is known among the Bedouins of Sinai. Drinking tea that contains a few leaf pieces of *H. boveanus* can be the *last tea party* for the consumer of this tea. *H. muticus* excels in the production of tropane alkaloids in its leaves: up to 6% (of dry weight) but the great majority of these alkaloids is hyoscyamine. It was originally intended to elevate the scopolamine/ hyoscyamine ratio. As Hashimoto *et al.* (1993), the Finnish investigators also used the gene for H6H and integrated it into the T-DNA of an *A. rhizogenes* vector and used the modified *A. rhizogenes* to obtain

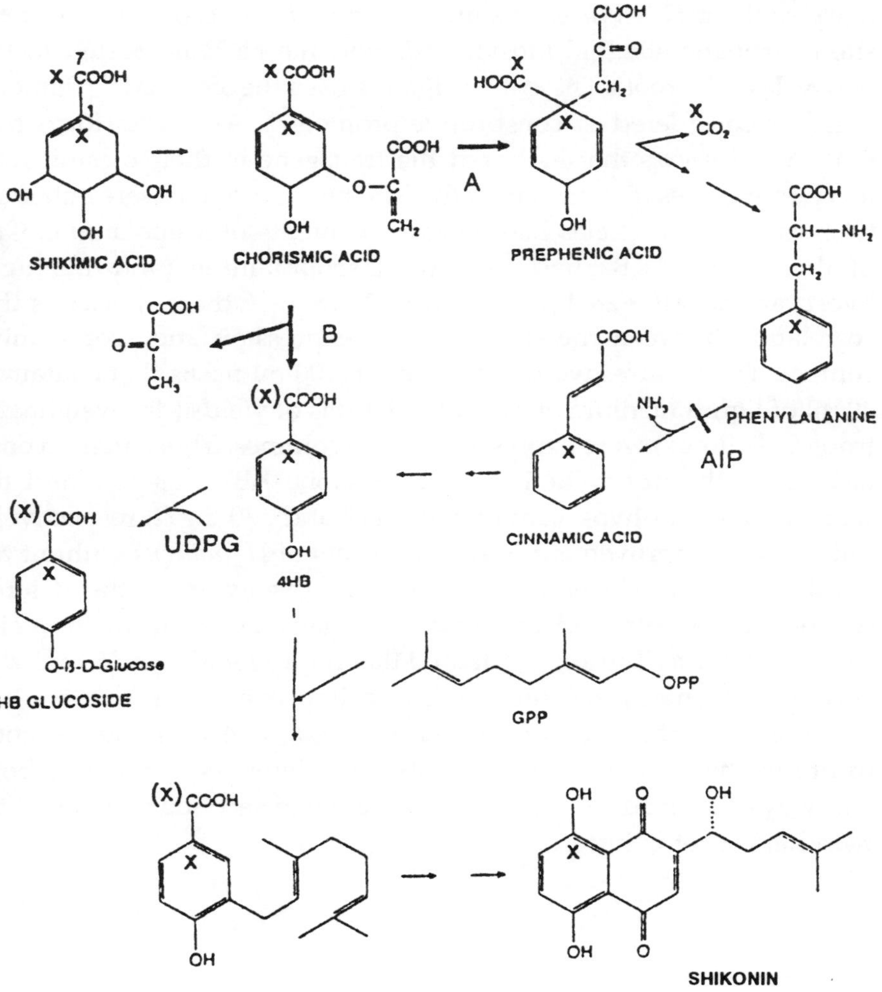

Fig. 28. Biosynthesis of 4-hydroxybenzoate and shikonin in *L. erythrorhizon*. (A) Pathway in untransformed cultures. (B) Pathway introduced by transformation with *ubiC*, the gene which encodes for chorismate pyruvate lyase. X shows the position of ^{13}C label in the feeding experiemnt with [1,7-$^{13}C_2$] shikimic acid. AIP, 2-aminoindan-2-phosphonic acid, an inhibitor of phenylalanine ammonia lyase; 4HB, 4-hydroxybenzoic acid; UDPG, uridine diphosphoglucose. (From Sommer *et al*., 1999.)

hairy roots of *H. muticus*. It should be noted that roots are the main site of tropane alkaloid biosynthesis and the H6H is located in the pericycle of the roots. These investigators used the 35S CaMV promoter which is considered a "constitutive promoter". After transformation with *A. rhizogenes* that harbored the transgene in their plasmids, 43 hairy-root clones that contained the 35S–h6h transgene were obtained. Of these, 22 root clones had elevated contents of scopolamine. One of these, clone KB7 had 14.4 mg/l scopolamine (and 125 mg/l hyoscyamine) after 28 days of culture. Moreover, there was a positive correlation between the level of H6H expression and scopolamine content. The cultures were performed in 100 ml flasks that contained 20 ml of liquid medium but the calculations of yields (dry weight and tropan alkaloids) were expressed in liter volumes. Thus, after 35 days of culture, the dry weight of roots in clone KB7 was 15 g and the scopolamine and hyoscyamine were 117 and 170 mg/l, respectively. This was an improvement over hairy roots of *H. muticus* without the addition of the *h6h* gene but still a long way from the desired scopolamine levels, and the yield is actually less than in the hairy roots of *Duboisia*. The authors hoped that improvements in knowledge and experimental procedures will also improve future results.

We share the hope of Jouhikainen *et al.* (1999) and see a good future in the combination of sensible interferences in the metabolic pathways that lead to production of valuable secondary metabolites with hairy-root cultures.

Chapter 6

General Considerations

6.1. Introduction

After we have surveyed the manufacture of medical and health products, we are now almost ready to ask some major questions. But before that, we shall summarize the major biotechnological achievements in this field of research. We shall see that these results indicate that in general it is feasible, although not devoid of difficulties, to produce reasonable quantities of active and safe biomedical products in transgenic plants. In subsequent sections of this chapter, we shall indicate what is still missing and attempt to predict the future development in this endeavor. We may thereafter ask: What are the problems involved in the production of therapeutics (and other health commodities) by genetically modified plants? What are the risks involved? What is the impact of public acceptance on research aimed to produce therapeutics in transgenic plants?

Among these problems, there is a group of issues that we shall not deal with. The reason for that is not because these problems are not important but rather that the authors of this book do not feel that they are competent to deal with them in an appropriate manner. We shall not deal with proprietary rights. Many techniques, plasmids and even genetic information procedures are "protected" by patents and so are specific bacterial lines, modified viruses and plant lines that are useful in the biotechnology of manufacturing therapeutics by transgenic plants. We provided an expert evaluation of this issue in a previous book coauthored by one of us (Galun and Breiman, 1997). The laws and regulations are changing constantly and this issue became very intricate. But, the issue is very important for those intending to be engaged in this biotechnology. Therefore, it is strongly recommended that experts are consulted at the very early stage of a study which intends to furnish useful products by transgenic plants.

The other topic with which we shall not deal is the economic-commercial issue. Obviously, it is a major issue if one aims to bring to the market a product that will result in commercial gains. In Chaps. 3–5, we mentioned several times the "calculations" made by the respective investigators who estimated the cost of manufacturing a given product versus its market price. In some cases, these calculations are obviously misleading for several reasons. A medical product may be replaced by another one and the price may change drastically to a fraction of the price cited in the calculation. Furthermore, the direct cost of manufacturing a product is only a minor component of the total cost. Those engaged in the biotechnology of producing health products know very well that having an apparently good idea constitutes a very early and risky beginning that goes a long way. In summary, this economic-commercial issue requires careful considerations and very rarely are the investigators aware of the pitfalls. In big corporations, the advisors on this issue are "inhouse", but it would be better for others to consult experts before they take on this risky and long journey. The following advice is attributed to the German statesman Otto von Bismarck (1815–1898) "Only the fools learn form their own experience". But one should take Bismarck's advice with a grain of salt: Bismarck himself was dismissed in 1890 by William II, the strong-minded emperor of Germany. On that, there is a saying (Mishna, Nezikin, Avot, Chap. 2) by Hillel the Elder who observed a scalp floating on the water; he said to the scalp: "Al deateft atifuch vesof metifaich yetufun" (the literal translation of this Arameic maxim is: you were punished for your wrongdoings and those who punished you will be punished).

6.2. Main Achievements

After reviewing the various medical and health products that were derived form transgenic plants, we can now look back at the state of the art. We shall do so according to the groups of products, meaning: antibody fragments and full-size antibody (see Chap. 3), antigens (see Chap. 4) and other medical and health products (see Chap. 5). We like to reiterate that we based our following summaries only on peer-reviewed publications. We avoided other publications such as patent applications, patents, commercial announcements and "submitted" manuscripts. The final test for the manufacture of therapeutic products is that they passed the due clinical tests, are produced commercially and the actual production costs are competitive to other means of production. Before we go into details, we can say that, with one or two exceptions, none of the transgenic-plant-derived products passed this final test. On the other hand, several products are rather close to this final test. We should not be surprised because the biotechnology of genetic transformation of plants emerged only about 15 years ago. Moreover, in some plants (such as cereals), efficient genetic transformation methods became available only recently and the ability to express alien genes in specific plant organs (e.g. grains of cereals and seeds of legumes) evolved only during the last few years. Also, in the early years of genetic transformation, investigators kept an attitude that turned out to be rather naive. This attitude assumed that once an alien gene is expressed in a plant cell, it is in its final form and stays as it is. This turned out to be completely wrong. For example, when the product is a polypeptide, it may have to be processed in various ways such as being folded correctly, glycosylated and "matured" by the removal of peptide sequences. In some cases, an elaborate trafficking in and out of subcellular organelles (e.g. the endoplasmic reticulum, the Golgi apparatus) is required, and finally secretion into the intercellular spaces. For these processings, it is essential to add components beyond the DNA sequence that codes for the required polypeptide. All these and other considerations emerged frequently through trial and error. Thus, while we emphasized (at the end of Chap. 2) that one should devote a lot of time to planning the work, especially during the construction of the transformation vectors, there is still a problem: in many cases, one does not know the obstacles beforehand. But even when the planning was meticulous and no major problems occurred along the way, it will still require several years before a good yield of products can be harvested from transgenic plants. Then, because we are dealing with medical products, they have to pass severe tests before they can be applied in human medicine. All these usually mean about ten years, from the time the investigator has a good idea until the product is ready for therapeutic application.

6.2.1. *Antibody fragments and full-size antibody*

6.2.1.1. *Antibody fragments*

The use of antibody in pathogenic conditions is mainly to neutralize directly the pathogenic entity (e.g. toxins, viruses, pathogenic bacteria). The antibodies provided during treatment are thus substituting the antibodies that are not produced (or produced but in a too low level) by the B-cells of the body. In cases where alien antibodies are required, they are needed instantly and in high quantity. This may constitute a problem: the applied antibodies should be free of any entities that will trigger adversely the immune system (e.g. if the antibodies are provided in horse serum, the patients immune system will be adversely affected and the patient will become extremely sensitive to any additional application of horse serum). In this respect, there is an advantage of plant-derived antibodies. They may be available in a more concentrated and pure form and free of human pathogens. Moreover, progress in immunology makes the production of potent antibodies by transgenic plants even more promising. As indicated in Chap. 3, it was found that for specific binding to an antigen, a fragment of the full-size antibody may suffice. This artificial fragment, termed single-chain variable fragment (scF_v), has only a fragment of the molecular mass of the full-size antibody (about 25 kDa) and it is composed of the variable domains of the light and heavy chains that are linked by a short chain of amino acids. The advantage of scF_v for therapeutic purposes is twofold. First, due to its much smaller size, this antibody fragment can reach the sites where it is required much more swiftly than the more bulky full-size antibody. The other advantage is in synthesizing the coding sequences for such antibody fragments. Once the amino acid sequences of the components of an effective scF_v are known, the respective cDNA can be synthesized. For expression in transgenic plants, this cDNA can then be integrated into the appropriate transformation vector. A stepwise progress in the utilization of scF_v in combination with genetic transformation of plants was described in some detail in Sec. 3.7. Thus, starting with Owen *et al.* (1992) who produced in transgenic plants scF_v that binds the plant photoreceptor phytochrome, through other studies by Firek *et al.* (1993), Tavladoraki *et al.* (1993), Artsaenko *et al.* (1995), Bruyns *et al.* (1996), De Jaeger *et al.* (1997) and Fischer *et al.* (1999d), it was revealed that transgenic plants can indeed be used for the production of potent scF_v with the expected binding properties. But, none of the above mentioned studies was directed to produce scF_v for use in human medicine.

A recent achievement with antibody fragments produced in plants concerns non-Hodgkins lymphoma (NHL) in human patients. McCormick *et al.* (1999) inserted the coding sequence of a specific scF_v into tobacco mosaic virus (TMV) in a way that after the plant was infected with the modified TMV, the scF_v appeared as epitopes on the coat protein of the virus. These scF_vs had an antitumor effect in a mouse model for NHL. Thus, transgenic plants are a promising candidate in contributing to the ongoing trend to furnish specific therapeutic antibody to individual patients.

6.2.1.2. *Full-size antibody*

The advantage of transgenic plants as producers of full-size antibodies has already been demonstrated by Hiatt *et al.* (1989). Two kinds of transgenic plants were produced. One group expressed only the heavy γ-chain while the other produced only the light κ-chain. When the two transgenic plants were cross-pollinated, some of the progeny plants produced the full-size antibody. In later studies, it was found that the coding sequences encoding the two chains can be engineered in the same transformation vector, each with its own promoter. Thus, the full-size antibodies can be assembled correctly in the plant cell. This assembly provided the possibility of obtaining chimeric antibodies in which the light and the heavy chains could be from different sources (e.g. Von Engelen *et al.*, 1994). Investigators also learned how to express the components of the antibody in specific organs of the plant (e.g. in seeds) or to cause the secretion of the antibody into the intercellular spaces. It also became evident that the correct processing, folding and assembly of functional antibody takes place in the endoplasmic reticulum. Therefore, an appropriate signal is required.

Another achievement was that functional antibody could also be produced in cultured hairy roots. For that, a two-step transformation was performed. First, transgenic plants were obtained (by Ma *et al.*, 1994) that produced Guy's 13 antibody (against a bacterium that causes teeth caries). These transgenic plants were then transformed with *Agrobacteirum rhizogenes* that induced hairy roots. Although this procedure still requires several changes in the methodology, the prospect that antibodies may be produced in hairy roots and secreted into the culture medium is very appealing.

The usefulness of transgenic-plant-derived antibodies (*plantibodies)* for therapeutic purposes was actually demonstrated earlier (Ma *et al.* 1994; 1998). Antibodies against *Streptococcus mutans* were applied to human volunteers and this application reduced bacterial colonization in the teeth.

Another research in this line is approaching applicability to patients. Zeitlin *et al.* (1998) used existing anti-herpes-simplex-virus (HSV-2) monoclonal antibody to derive the respective codes for such antibodies. These codes were inserted into transformation cassettes and after transformation of soybean plants, the respective plantibodies were obtained. These antibodies were able to neutralize HSV-2 and the plantibodies were stable in the human vagina for a day or longer.

6.2.2. *Antigens*

The first achievement in the field of producing therapeutic antigens in transgenic plants was reported by Mason *et al.* (1992). This was followed by the Arntzen–Mason team (Thanavala *et al.*, 1995). These investigators expressed a recombinant hepatitis B surface antigen (rHBSAg) in transgenic tobacco plants. Mouse T-cells could be stimulated by repeated injection of rHBsAg. However, the levels of rHBsAg in tobacco leaves was rather low. The system was not carried further or the progress was not yet reported in reviewed publications. Also, attempts by another team to express a HBV antigen in potato did not lead to high levels of antigen (Domansky *et al.*, 1995; Ehsani *et al.*, 1997).

An attempt to produce an edible antigen against rabies virus in tomato also resulted in only very low antigen levels (McGravey *et al.*, 1995). Work on the production of antigens against the Norwalk virus (NV) for oral vaccination (Mason *et al.*, 1996; Ball *et al.* 1999) reached the state whereby transformation of tobacco and potato plants led to the production of the coat protein of NV with its assembly into virus-like particles. These resembled morphologically the authentic NV particles. Gastric application of the plant-derived antigen to mice elicited the respective IgG and IgA responses. Moreover, the plant-derived antigen did not cause deleterious effect in human volunteers. But, these studies did not reach the crucial stage of neutralizing activity against NV. A similar stage was reached with the swine-transmissible gastroenteritis virus (TGEV) by Gomez *et al.* (1998). In this case, the bottleneck was the low level of TGEV antigen produced in the transgenic plants.

One further step in the production of an antiviral vaccine in plants was achieved for the foot and mouth disease virus (FMDV). In this viral disease (Corrillo *et al.*, 1998; Wigdorowitz *et al.*, 1999), the research reached the stage whereby mice that were immunized with the plant-derived antigen were found to be protected against a subsequent challenge with FMDV.

The research that aimed to produce antigens for oral immunization against bacterial diseases reached a fairly advanced stage, albeit it did not attain any medical-therapeutic application yet. The diseases caused by *Vibrio cholerae* and enterotoxigenic *E. coli* (ETEC) were handled by several studies that aimed to produce effective oral vaccinations against these diseases (Haq *et al.*, 1995; Arakawa *et al.*, 1997; 1998a; 1998b; Mason *et al.*, 1998; Tacket *et al.*, 1998). In both *V. cholerae* and ETEC, the toxic subunit A is attached to the G_{m1} ganglioside by subunit B. If the binding of the subunit B is eliminated, there is no toxic effect. Thus, eliciting specific antibody that will interfere with the binding of subunit B will actually eliminate these diseases. Antigens that elicited such antibody in mice were produced by transgenic plants, and the immunized mice showed reduced diarrhea when subsequently infected with cholera. The antigens inside transgenic potato tubers (not cooked) were provided orally to volunteers without ill effects. Most of these volunteers had an increase of the required antibody. But the studies did not yet reach the stage of large-scale medical tests.

There was considerable success in the production of a fusion protein in transgenic potato tubers that could be useful to counteract insulin-dependent diabetes mellitus (IDDM) — an autoimmune disease. Arakawa *et al.* (1998b) constructed a transformation plasmid that contained the code for a fusion protein: CT-B/insulin conjugate. Potato tubers that expressed this fusion protein reduced the incidence of IDDM in five weeks old NOD mice fed repeatedly with these tubers. But only future research will tell whether or not this idea can be useful in reducing human IDDM.

Impressive biotechnological progress was achieved in recruiting plant viruses to express antigens. The "beauty" of this approach is that plants systemically infected with such a virus will contain a high level of coat protein (CP) of this virus in all their leaves. The virus CP may reach about 25% of the leaf proteins, about three weeks after infection. Several groups of investigators (Donson *et al.*, 1991; Hamamoto *et al.*, 1993; Usha *et al.*, 1993; Porta *et al.*, 1994; Fischer *et al.*, 1995; Porta and Lomonossoff, 1996; Dalsgaard *et al.*, 1997; Durrani *et al.*, 1998; Bendahmane *et al.*, 1999; Lomonossoff and Hamilton, 1999) contributed to the application of modified plant viruses (TMV and CPMV) with this approach. The overall idea was to modify the viral genome and to insert into it a sequence that will code for an antigen. The insertion was done in a way that will cause the expression of the antigen on the surface of the CP. In a recent study (Koo *et al.*, 1999), in which the system was improved in several criteria, it was found that mice immunized with an antigen derived from the modified TMV infected plants were protected against murine hepatitis virus (MHV). MHV is a major disease in facilities for the breeding and supply of experimental animals. Thus, an effective plant-derived vaccine against MHV would be of considerable economic importance.

Finally, with respect to antigens carried as epitopes on plant viruses, there was success in the production of antigens that activate adaptive immunity against diseases caused by the bacteria *Staphylococcus aureus* and *Pseudomonas aeruginosa*. These antigens were produced as epitopes on CPMV or TMV and were effective in mice models (e.g. Brennan *et al.*, 1999a; 1999b; 1999c; Staczek *et al.*, 2000). However, the studies only reached the phase whereby they are good *bases* for the development of vaccines for human patients.

6.2.3. *Therapeutic products that are unrelated to the immune system*

The manufacture of these products in transgenic plants, and in a few cases in transgenic plant cell cultures, was reviewed in Chap. 5. Here, we shall highlight the main achievements. We shall do it in the same order as presented in Chap. 5: transformation by plant viruses, transformation of cell suspensions, *Agrobacterium tumefaciens* transformation, biolistic transformation and *A. rhizogenes* transformation that resulted in hairy-root cultures.

6.2.3.1. *Transformation by plant virus infection*

This method of transformation was attempted in only a few cases and the yield of the product was not impressive in any of them.

One attempt was to produce the angiotensin-I-converting enzyme inhibitor (ACEI) in TMV infected plants (Hamamoto *et al.*, 1993). The infected plant was either tobacco or tomato. The coding sequence for the 12 amino acids that constitute the ACEI was placed downstream of the coat protein (CP) coding sequence and a six nucleotide sequence between the two codes allowed a partial readthrough. The level of CP-ACEI in the tobacco leaves reached 100 μg per g fresh weight and in tomato fruits, it reached 10 μg per g fresh weight. It should be noted that ACEI constituted only a small fraction of the CP-ACEI.

A somewhat higher level was obtained in a similar procedure to derive α-trichosanthin (an inhibitor of protein synthesis and a possible therapeutic agent against HIV). The α-trichosanthin reached up to 2% of the soluble protein in modified TMV infected tobacco leaves (Kumagai *et al.*, 1993).

6.2.3.2. *Transformation of cell suspensions*

This procedure of transformation could be applied either to cell cultures that maintained the capability to redifferentiate into functional cells (e.g. rice embryogenic cell cultures) or to cell cultures that lost this capability (e.g. the BY-2 cell line of tobacco). In the latter case, the cell cultures can be cloned as callus lines but the presently available biotechnology cannot regenerate functional plants from these calli.

The rice system was used in a study that was aimed to produce interferon-αD (Zhu *et al.*, 1993). The interferon was found in rice cell lines and in rice plants that were regenerated from protoplasts derived from such cell lines. But, apparently, the work did not reach the real production level.

Tobacco cell cultures (BY-2) were transformed to express human erythropoietin (Epo) (Matsumoto *et al.*, 1995). The activity of BY-2-derived Epo was lower than expected. This possibly resulted from the glycosylation in the plant cells which is different from the glycosylation in mammals. In a similar work, Magnuson *et al.* (1998) produced interleukin-2 and interleukin-4 in tobacco cells. As found with Epo production, the level of interleukins, evaluated by ELISA tests, was also much higher than the level evaluated from the activity of the plant-cell-derived interleukins. We did not trace a continuation of this work, but as only about two years have lapsed since the published work, further results may be forthcoming.

6.2.3.3. *Agrobacterium tumefaciens-mediated transformation*

This transformation was used in numerous cases in which attempts were made to produce therapeutic peptides and proteins in transgenic plants. One of the earliest attempts was also one of the most successful (Vandekerchkhove *et al.*, 1989). The researchers involved should be credited for their sensible strategy. They chose the correct site of expression (seeds) and accordingly engineered the transformation vector. The code for the pentapeptide that constitutes leu-eukephalin was inserted into the code for a seed protein and *Arabidopsis* plants were respectively transformed with an *Agrobacterium* strain that harbored the engineered vector. The pentapeptide was thus accumulated in the protein bodies of the *Arabidopsis* seeds. Was this system moved from *Arabidopsis* to a related crop plant (rape) and was further development of this work moved from the Department of Genetics of the University of Gent to the nearby commercial company Advanced Genetics Systems that had close relations with this department? We do not know; what we do know is that the company went through a major transition.

Work with P-450, protoheme-containing monooxygenases, indicated an interesting aspect of the plant-derived medical products: the product may strongly affect the transgenic plant in which it is produced. This aspect became apparent from the work of Saito *et al.*

(1991) who intended to produce the rabbit liver cytrochrome P-450 in transgenic tobacco plants. The investigators observed, right after the due *Agrobacterium*-mediated transformation, that when they used a plasmid which contained the coding sequence for this product, the regeneration of the transgenic tobacco plants was sluggish. When, after considerable delay, plants were obtained, they flowered and senescenced early. They were also modified in other ways. The investigators concluded that the mammalian P-450 adversely affected the plants. Thus, one should take such a possibility into account: the expression of a mammalian enzyme, that has a parallel in plants, may affect the metabolism of the plants.

Another important aspect was revealed when an attempt was made to express the trout growth hormone (tGH-II) in transgenic plants. Bosch *et al.* (1994) found that when the transformation plasmid was engineered for expression at the correct sites (e.g. leaves of tobacco or seeds of *Arabidopsis)* and after transformation resulted in apparently correct processing and folding, the final product still lacked biological activity. Do slight changes in glycosylation reduce biological activity? We do not know yet but there is a warning: the biochemical detection of a product does not assure that a biologically active substance was produced.

Obviously, no such problems are expected when a product that is normally produced in a plant is to be expressed in another (transgenic) plant. This was shown for the very potent toxin, ricin. In a similar manner to other toxins (e.g. the cholera toxin), ricin is composed of two polypeptide chains: RTA and RTB. The RTA has a strong ribosome-inactivating activity while the RTB has a galactose-binding activity. The latter chain thus binds the ricin to the target cell. If the RTA is to be used for ablation of specific cells, it should be free of RTB. Investigations by Sehnke *et al.* (1994), and Sehnke and Ferl (1999) showed that ricin can be produced in transgenic tobacco plants and cell cultures derived from such plants, and ricin was then released into the culture medium, reaching 25 ng/ml. The recombinant ricin was identical to the authentic ricin and could be separated into

RTA and RTB chains. The tobacco-derived ricin was as active biologically as the castor-bean ricin.

A successful endeavor was the production of biologically active hirudin in the oil bodies of transgenic rape seed (Parmenter *et al.*, 1995). Hirudin is a small protein (7 kDa) produced by leeches. Correct engineering of a transformation vector followed by *Agrobacterium*-mediated transformation resulted in rapeseeds that contained hirudin in their oleosin. Cleavage of the oleosin–hirudin fusion protein followed by due separation resulted in pure hirudin that had the expected antithrombin activity. This does not mean that the medical future of leeches is threatened because the saliva of leeches contains several components in addition to hirudin that still render leeches useful for specific medical purposes.

Our last example of a successful manufacture of an alien compound in transgenic plants concerns a health product rather than a therapeutic product. Ye *et al.* (2000) succeeded in obtaining transgenic rice that produced grains with β-carotene in their endosperm. The level of β-carotene was high enough to cause the rice grains to be pigmented. Whether or not this staple food with bacterial enzymes will be accepted by the public is an open question but the biotechnological achievement is impressive.

6.2.3.4. *Biolistic transformation*

With this method of genetic transformation of plants, a product was obtained that reached the commercial market: avidin. The work was performed by Hood *et al.* (1997, 1999). Avidin is a protein composed of four identical subunits of 128 amino acids each. It is used as a tool in molecular biology and may also have therapeutic uses. The previous source of avidin was chicken eggs. Hood and collaborators designed the required transformation vectors with coding sequences that were modified to conform with codon preference of maize and used these vectors in biolistic transformation of maize. The avidin could be isolated from the embryos of transgenic maize grains and the yield reached 230 mg/kg kernels. This recombinant avidin is now commercially produced.

A similar study to that for avidin production was conducted in a collaborative effort between the avidin team and others (Zhong *et al.*, 1999). It also involved maize transformation but was aimed to derive another product: aprotinin. This is a serine proteinase inhibitor used in biochemical research and as a therapeutic agent. These investigators succeeded in producing high levels of aprotinin in the embryos of maize kernels. The recombinant aprotinin was stable in the kernels and had the same biological activity as authentic aprotinin.

6.2.3.5. *Agrobacterium rhizogenes transformation*

The studies that aimed to derive valuable therapeutic products from plants transformed by *A. rhizogenes* can be divided into two main research lines. In one research line, plants that were known to contain a therapeutic compound were induced to produce hairy roots by this transformation. The rationale behind this line of research was that the established culture of hairy root will then constitute a good source of the respective compound. The other line of research was to interfere with the secondary metabolism in the hairy roots by including in the *A. rhizogenes* a plasmid with a coding sequence of an enzyme that will shift the metabolism in such a way that more of the required compound will be produced in the hairy roots.

In the first line of research, no extraordinary biotechnological results were to be expected because the transformation was not intended to cause the expression of a transgene. But the results of a great number of studies indicated that many medicinal plants can be induced to produce hairy-root cultures and that these cultures, in most cases, indeed produced the therapeutic compounds. The compound either stayed in the roots or in a few cases (such as *Lithospermum erythrorhizon* roots producing shikonin, see Shimomura *et al.*, 1991b), the compound was (also) released in great quantities to the culture medium from which it could be extracted continuously. A good example for a therapeutic compound produced in hairy roots is artemisinin, an anti-malarial agent. Liu *et al.* (1998) were able to obtain about 540 mg of artemisinin per liter of hairy roots of *Artemisia annua* from 2-liter bioreactors.

One of the advantages of producing such compounds in cultured hairy roots rather than in intact plants is the possibility to interfere in the production. This could be done by precursor feeding or other addition to the culture medium in order to shift the production into a desired pathway (see Morgan and Shanks, 2000, for literature).

Noteworthy achievements were recorded in the second research line in which the cDNAs for enzymes that affect the synthesis of the secondary metabolites were engineered into the *A. rhizogenes* plasmid. In this way, the metabolism of the hairy roots was changed. Thus, the scopalamine content of the hairy roots of *Atropa belladonna* (Yun *et al.*, 1992) and of *Hyoscyamus muticus* (Jouhikainen *et al.*, 1999) could be enhanced by the integration of the cDNA for the enzyme hyoscyamine-6β-hydroxylase into the respective genomes of the *Atropa* and *Hyoscyamus* species. However, the commercial value of scopolamine is not very high; therefore, commercial companies will have to decide whether to continue extracting scopolamine from intact plants (e.g. *Datura sanguinea*) or to shift the production to genetically modified hairy-root cultures.

6.3. What is Still Missing and What is Desirable in the Future

With only very few exceptions, none of the studies that aimed to manufacture therapeutic or health products reached the destined goal. Some studies are probably being continued and may reach their goal in the future. Other studies were possibly discontinued or went "underground". By "underground", we mean that they are performed without bringing the results to the public knowledge. By bringing the results to public knowledge, we mean publishing in peer-reviewed journals. We can only guess why the studies were discontinued. There could be at least four reasons for discontinuation:

(a) the product was found to lack biological activity (e.g. problems of processing of the transproteins);

(b) the level of product in the transgenic plants was found to be too low to justify the manufacture of the specific product in transgenic plants;

(c) the medical/financial value of the product declined and therefore there was no incentive to invest resources in further research and development;

(d) the investigators (or the respective sponsors of the research) ran out of financial resources.

There could be any combination of these reasons. For reasons (c) and (d), these are outside our realm. For our deliberations, reasons (a) and (b) indicate what is still missing.

We indicated already that out of the many examples of therapeutics derived from transgenic plants described in this book, only very few had matured to the stage of real critical assessment and only one product is in the market. Although the field of using transgenic plants for the manufacture of medical products is young, the fact that only one product is in the market warrants a short discussion to assess the causes for the relatively poor results.

The most critical step in developing a transgenic plant is not related to the methodology but to which target the research is aimed. As in navigation, the first course one takes may cause the most serious mistake. In the case of transgenic plants, the selection of the proper target is the most important decision in such a project. The general guidelines for the selection of a proper target are the following:

(a) the protein or the peptide that will be produced is for real clinical need and, if possible, used for more than one indication e.g. erythropoietin for the treatment of β-thalassemia and renal-failure-related anemia;

(b) the product could be consumed as for oral tolerance, vaccinated or applied topically;

(c) the product should tolerate minor N-terminal and C-terminal amino acid modifications;

(d) its production in bacteria or yeast is hampered, generating low yield or an inactive product resulting from abnormal folding or glycosylation;

(e) the therapeutic effect of the product is measurable, e.g. generation of neutralizing antibodies.

One of the clear examples in which investigators entered a very difficult situation in the process of translating their transgene products into clinical use is the production of antibodies. Previous publications had suggested to producing human, not humanized, monoclonal antibodies in plants. In such an example, it might be very difficult to generate in a plant cell the whole antibody in the right tertiary structure and carbohydrate modification, including its F_c and F_{ab} components. However, single-chain antibody harboring the desired targeting or antimicrobial properties may well be produced in transgenic plants. In this case, a single relatively short open reading frame could be constructed into a transgenic plant delivery cassette in a simple way. The biological properties of the product can then be easily measured and the therapeutic effects, in specific cases, could be evaluated.

Constant improvement in the understanding of protein processing, glycosylation and trafficking in plants are expected to be of great help in the future manufacture of medical products in plants, that will have biological activity which approaches the activity of the natural products. Such processes are not posing a problem for secondary metabolites that are usually relatively simple molecules. Thus, what is still missing in many cases where the product is a protein or a shorter peptide is a full understanding of the processing and trafficking that are required in order to cause similar processing and trafficking in the transgenic plants.

With respect to the yield of the product in transgenic plants, there are several issues that should be taken care of. One major issue is "silencing" of the transgene. This imposes a problem especially in two cases. One is if the transformation is not by agrobacteria. In such "direct" transformations, as biolistics, the transgene may be integrated in many chromosomal sites and this frequently causes mutual silencing. Recent studies (Srivastaba *et al.*, 1999) may furnish a solution by elimination of these multiple integrations. Another more "primitive" solution means more work... to produce a great number of transgenic plants in order to get a few transformants that are not silenced.

Silencing can also be caused when two transgenic plants have to be crossed in order to obtain, among the sexual progeny, plants that contain more than one transgene that will be expressed. In this case, it appears that silencing will be more frequent when the crossed plants share the same promoter. Using MARS (SARS) that will flank the transgene could reduce the silencing phenomenon (e.g. Han *et al.*, 1997; Gindullis and Meier, 1999; and see Chap. 1). Recent reviews on silencing and other mechanisms that repress gene expression handle this issue thoroughly (e.g. Laherty *et al.*, 1998; Baumann *et al.*, 1999; Bird and Wolffe, 1999; De Neve *et al.*, 1999; Gutierrez *et al.*, 1999; Knoepfler and Eisenman, 1999; Maldonado *et al.*, 1999; Sunand Elgin, 1999; Waterhouse *et al.*, 1999). It is therefore highly recommended that investigators who intend to launch a project of manufacturing a medical product in transgenic plants should update themselves in this issue before the start of a detailed planning of their research.

The capacity of the cytosol to store polypeptides and other medical products is limited. Therefore, future research should direct the accumulation of these products in other locations, such as seed storage bodies, intercellular spaces, chloroplasts, and plant virus coat proteins. When the production (e.g. of secondary metabolites) is to be in hairy roots, it may be preferable to secrete the products into the culture medium from where they could be continuously removed by a bypass collector. In all these cases, the appropriate coding sequences for sorting signals should be included in the transformation vector.

Further improvement in genetic transformation of crop plants, such as rape, soybeans, wheat, rice and maize will facilitate the work aimed at accumulating pharmaceutics in the storage tissue of such plants (e.g. oil bodies, aleurones).

6.4. Safety Issues

There are three safety issues that are relevant to the manufacture of medical and health products in transgenic plants. One safety issue concerns the laboratory work during the genetic engineering of the plasmids, their transfer among bacterial hosts and the transformation of the plants. The other issue concerns the environment in which the transgenic plants are grown and the further handling of these plants. The third issue concerns the safety of the therapeutic product.

In the USA and in countries that do not have laws and guidelines for handling the safety of transgenic plants, it is advisable to follow the USA laws, regulations and policies. These can be obtained from the website http://www.aphis.usda.gov/biotechnology/laws.html where there is an update of these laws and regulations.

6.4.1. *Laboratory work and culture room*

The safety concerns during laboratory work, such as the handling of DNA manipulations, radiolabeling, handling bacteria etc., in work aimed to manufacture therapeutics in transgenic plants are no different from those in any other molecular genetic research. These can be found in the respective handbooks that provide safety instructions for such work (e.g. Flaming *et al.*, 1995; Ausubel *et al.*, 1999) The main "tool" in plant genetic transformation is the use of agrobacteria; these post no threat to the person who handles them. They also do not constitute a threat to plants because the strains of *Agrobacterium* used in transformation are usually "disarmed": they lack the capability to parasitize plants. Even after the plasmid containing the borders of the T-DNA is introduced into such agrobacteria, the chances that they will "leak" out of the lab and cause a phytopathological danger is very close to nil. The several devices employed to perform the biolistic transformation should be used as instructed by the manufacturers but they do not involve dangers if handled properly. There is one warning that concerns a minority of the investigators. This warning relates to those who are extremely sensitive to one of the antibiotics used during cloning and selection of transgenic transformants (this is a personal experience of one of the authors of this book). Even the handling of rooted shoots that grow in a high concentration of an antibiotic compound without gloves may cause severe allergic effects on the fingers.

6.4.2. *Safety to the environment*

The transfer of transgenic plants from the culture room should be to a confined greenhouse that will bar the dispersal of pollen and seeds to the environment. The same precautions that are required for all transgenic plants are also required for transgenic plants that are used for the production of therapeutics. From the confined greenhouse, the transgenic plants may be destined to one of the following routes. Seeds and other propagation material may be returned to the laboratory for further handling; plant tissue will be collected for analyses in the laboratory; all the remaining plant material should be incinerated. If large-scale culture of the transgenic plants that produce pharmaceutics is required, field experiments (or commercial fields) should be performed according to the license requirements in each country.

For the safety of the public, it is highly recommended that all the operations that involve the production and maintenance of transgenic plants for therapeutics and health products, be performed in a special facility for transgenic plants. This facility should prevent access to unauthorized people.

The above noted paragraphs indicate that the environmental safety issues with respect to transgenic plants for the production of therapeutics, do not differ from environmental safety issues concerning other transgenic plants (e.g. for crop improvement). The environmental safety issues with the latter transgenic plants were amply discussed in recent years. For information on these discussions, the interested reader may consult Chamberlain and Steward (1999), Raybould (1999), and Gressel and Rotteveel (2000). But, we should remember that the fields of transgenic plants for the manufacture of therapeutics (that presently do not exist) will be by far smaller than fields of transgenic plants for food production.

Harry Daniell (1999) came up with an idea to assure environmental safety that he called: "Environmentally friendly approaches to genetic engineering". Actually, Daniell dealt with the problem of dispersal, by pollen, of genes from transgenic crop plants to other plants (e.g. wild relative). His remedy was to introduce the transgenes into the plastome (the chloroplast genome) rather than into the nuclear genome. The argument of Daniell was that in crop plants, the pollen does not transmit plastomic genes. Well, it is a nice idea but there are two problems. First, in no case is paternal transmission of chloroplast genes eliminated completely. Even in tobacco where there seemed to be "absolute" maternal transmission of plastomic genes, paternal transmission was observed after selection. Thus, if such transmission is even only 0.01%, then 1/10,000 of the progeny that result from a cross of pollen from the transgenic cultivar to a wild relative will contain the plastomic transgene. Under due selective conditions (when the plastomic gene confers herbicide resistance and this progeny grows in a herbicide-treated field), there is a fair chance that the transgene will be introgressed into the wild species. Actually, there is no reliable information on the level of paternal transmission of plastomic genes in crop plants such as rice.

The other problem with maternal transmission of transgenes is that this transmission does not permit the recombination of two transgenes from two different transgenic plants into one plant by crossing and selection. Thus, Gressel (1999) suggested another approach that could reduce the danger of drift from transgenic cultivars to their wild-related weeds. In summary, one should be careful with respect to pollen dispersal. For example, if a field of transgenic maize plants that harbor a therapeutic product is adjacent to another maize field, the air-borne pollen from the transgenic maize may transfer the transgene.

This awareness to environmental safety became apparent in a recent publication by the Monsanto Company, a company that was previously the target of criticism by environmentalists. In a very elegant study (Staub *et al.*, 2000), the Monsanto team combined awareness of pollen dispersal from transgenic plants with several other biotechnological considerations. They used the basal approach of Maliga and collaborators (see Sec. 2.7) to utilize the homologous recombination in the plastome and selection on spectinomycin to introduce a gene encoding the mature human somatotropin (hST) into the plastome. The hST has several uses in therapeutic medicine as in the treatment of hypopituitary dwarfism in children. A rather complicated and cleverly engineered plasmid was used in the biolistic transformation of tobacco. Considerable quantities of chemically authentic hST were obtained in these transplastomic tobacco plants. The biological activity was verified *in vitro* by the assay of enhanced proliferation of rat lymphoma cells.

6.4.3. *Safety of products*

Once peptides or proteins are expressed and produced from transgenic plants, it is essential to evaluate their activity and safety. This issue is particularly important if the products are to be introduced into human or animals parenterally. For each specific case, a side-by-side *in vivo* evaluation of toxicity and efficacy in a dose escalation study is highly recommended.

There is a variety of structural differences that could occur while producing therapeutic polypeptides in the plant cellular environment.

Protein folding in plant cells has more similarities to the folding of proteins in mammalian cells than in yeast or bacteria. Usually, aberrant folding of proteins will not induce a toxic effect but rather a reduced one; however, each specific case should be assessed. There is therefore an advantage of choosing as a goal a therapeutic protein that is not glycosylated. This is, for example, the case in human somatotropin (Staub *et al.*, 2000). The potential structural differences in proteins produced in plants as compared to mammalian cells are at the post-translational phase of protein processing:

(a) Several enzymes are anchored to cell membranes through glycosylphosphatidyl inositol. One important example is the enzyme alkaline phosphatase. Phospholipid anchoring is essential for the enzymatic activity of this enzyme. The expression of proteins in plant cells could prevent the proper cellular localization of this protein to exert its activity.

(b) Myristoylation, palmitoylation and the addition of a farnesyl moiety to proteins enable proteins to be localized in the inner surface of the cell membrane. Furthermore, some of these modifications are responsible for protein–protein interactions and subcellular compartmentization. Lipophilization of proteins could also improve effects of peptides or proteins as has recently been reported for the somatostatin analog RC-160 (Dasgupta and Mukherjee, 2000). Such modifications could not be introduced into polypeptides in plant cells in their natural environment with additional transgene modifications.

(c) The most important post-translational modification, related to a vast majority of proteins, is the covalent binding of carbohydrates. Specific differences in the glycosylation of polyproteins could alter their effects. The glycosylation of the F_c portion of antibodies is essential for F_c receptor binding and the induction of antibody-dependent cell-mediated cytotoxicity (ADCC). Glycosylation of the hinge region of antibodies prolongs the half-life of this protein in the serum. In addition to the pharmacokinetic differences related to variance in plant-derived protein glycosylation, the activity, affinity — in the case of antibodies, and immunogenicity could be present, following the production of the proteins.

In an effort to overcome the major limitations related to post-translational modification, a panel of chemical, physical, pharmacological, biological and immunological tests should be conducted in each particular case. Some of these tests could be conducted *in vitro* for the assessment of biological activity. However, no conclusive result could be taken prior to a rigorous analysis of each particular product *in vivo* in the most relevant animal model. In specific examples, it might be impossible to conduct animal experiments. Again, a side-by-side assessment of the plant and mammalian products of any polypeptide is crucial before any Phase I study in humans takes place.

6.5. Public Acceptance

The question of public acceptance of genetically modified (GM) food attracted great attention in recent years. The public acceptance of therapeutics derived from transgenic plants did not gain such attention. One possible reason for that may be that such products did not reach the market. There are products that border between food and therapeutics, such as β-carotene and other health products. The issue of public acceptance may be raised in the future, at least for these "border" products. Because the present discussion of public acceptance concerns GM food, we shall not review this discussion here. The issue was discussed very thoroughly by Sasson (1999) and in several recent meetings (e.g. Raybould, 1999). The trend to avoid GM food and the rather aggressive opposition even to grow transgenic plants that were intended to serve as food had a major impact in many developed countries. This avoidance and aggressive opposition adversely affected major commercial companies that invested huge resources to develop such transgenic plants.

We assume that at least in part the reluctance of a population to consume GM food is not based on rational arguments but on emotions and on lack of information. The latter possibility is highlighted by a pull in a developed European country, according to which 80% of the people interviewed claimed that they would never consume DNA. Dealing rationally with irrational issues is difficult. The late Professor Gershon Sholem, the "father" of academic research on Jewish mysticism was once invited to lecture at the Jewish Theological Seminary (an institute of the Conservative Jewish Movement) in New York. After the lecture, the head of the seminary summarized: "... You see, nonsense is always nonsense, but the science of nonsense is a science". Sholem was deeply insulted.

But, we shall take an approach that completely differs from the one taken by some experts in the field. We do not intend to change the attitude of the public. Neither will we ask the philosophical question of whether or not ignorant people should participate in the decision of an issue in which they are not informed. We shall also not attempt in this book to educate the public nor shall we question the wisdom of public emotions. Our approach is that we shall take the opinion of the public as it is. On the other hand, we strongly recommend that the work which aims at the manufacture of medical and health products will be performed in a way that there should not be any rational reason for public opposition. Also, we pointed out above that the goals of such a work should be restricted to products of medical relevance. People are aware of the fact that many therapeutics are not only *unhealthy*, they can be extremely toxic (e.g. taxol and other poisonous compounds that kill cells). If such therapeutics are produced by transgenic plants rather than being extracted from "natural" plants, there is no rational reason to avoid genetic transformation. The same logic can be applied to other therapeutics such as antibodies that suppress specifically cancer cells.

There is again a borderline. One can genetically modify the lipids in oilseeds, rendering the oil to a better and possibly more healthy quality. Such borderline goals are problematic and we avoided them in this book.

6.6. Epilogue

The Preface of this book provided some reflections on features of human nature that are related to the investigative urge. These reflections are not only relevant to any human endeavor that is aimed at revealing new information but they are also especially applicable to scientific efforts where there is a combination of new and emerging methodologies and ideas that may provide results which are of real benefit to man. Human nature and man's inclination to embark on the risky path of scientific investigation is an interesting and well-researched subject. We can begin to dissect human nature in this respect on a very sound basis. This basis was formulated by David Hume (1711–1776) in "*A Treatise of the Human Nature*" (1739), a philosophical contemplation that is still valid today. In the epilogue of this treatise, Hume confessed that his urge to contribute to the understanding of the "human nature" was driven by his *lust* to acquire *fame* in this field of knowledge. We may add to Hume's deliberations another very important instinct that man shares with other animals, but is especially strong in humans: curiosity. We could thus discuss emotions and instincts that drive man to risky endeavors. However, this interesting subject is outside the realm of our book and the authors do not intend to delve into it. We shall stay within the rational realm. When we surveyed publications on the manufacture of therapeutics by genetically modified (GM) plants, we noted in several cases the claim that GM plants are a preferred source for such therapeutics. A number of reasons were provided for this argument, such as: therapeutics derived from transgenic plants will be cheaper or the product will be free of contamination (e.g. by pathogenic human viruses). Do we have evidence that these claims are true? There are only *very* few cases in which a useful therapeutic product has been proven and we are not aware of any case in which such a product has been authorized for use in medical practice (an exception is avidin but this is not a therapeutic product although it is useful in biomedical research and is therefore included in our book). On the other hand, there are quite a number of cases where a good idea, meticulous planning and professional execution have been combined to yield something that is rather close to a useful transgenic plant product. Can we say that, out of an emotional drive supported by rational thinking, there is a reasonable future for this field of endeavor? We tend to answer in the affirmative.

There is an additional issue that should not be neglected and this is the financial-economic one. The time-span from a sound idea to the commercialization of a therapeutic product is very long. This time-span is estimated to be between 8 and 15 years. Such a long time carries with it two problems: financing and development. We have witnessed several commercial laboratories exhausting all their financial resources before they could recoup from their achievements. Then again, a therapeutic product may lose its commercial value over the long development years because it is replaced by another product.

However, in spite of all reservations, we tend to trust in human genius. While Immanuel Kant has claimed that: *"Out of the crooked timber of humanity no straight thing was ever made"*, we still have confidence that the human mind will find remedies for its body's maladies and that some of these remedies will come from transgenic plants.

References

Alberts, B., Bray, D., Lewis, J., Raff, M., Roberts, K. and Watson, J.D. (1994). *The Molecular Biology of the Cell* (3rd edition), Garland Publishing, NY, 1294 pp.

Aoki, T., Matsumoto, H., Asako, Y., Matsunaga, Y. and Shimomura, K. (1997). Variation of alkaloid productivity among several clones of hairy roots and regenerated plants of *Atropa belladonna* transformed with *Agrobacterium rhizogenes* 15834. *Plant Cell Rep.* **16**: 282–286.

Arakawa, T., Chong, D.K.X. and Langridge, W.H.R. (1998a). Efficacy of a food-plant-based oral cholera toxin B subunit vaccine. *Nature Biotech.* **16**: 292–297.

Arakawa, T., Chong, D.K.X., Merritt, J.L. and Langridge, W.H.R. (1997). Expression of cholera toxin B subunit oligomers in transgenic potato plants. *Transgenic Res.* **6**: 403–413.

Arakawa, T., Yu, J., Chong, D.K.X., Hough, J., Engen, P.C. and Langridge, W.H.R. (1998b). A plant-based cholera toxin B subunit–insulin fusion protein protects against the development of autoimmune diabetes. *Nature Biotech.* **16**: 934–938.

Arntzen, C.J. (1998). Pharmaceutical foodstuffs — Oral immunization with transgenic plants. *Nature Medicine* **4**: 502–503.

Artsaenko, O., Peisker, M., Nieden, U.Z., Fiedler, U., Weiler, E.W., Müntz, K. and Conrad, U. (1995). Expression of a single-chain F_v

antibody against abscisic acid creates a wilty phenotype in transgenic tobacco. *Plant J.* **8**: 745–750.

Ausubel, F.M., Brent, R., Kingston, R.E., *et al.* (eds.) (1999). *Current Protocols in Molecular Biology*, John Wiley & Sons, Inc., NY.

Avery, O.T., McCleod, C.M. and McCarthy, M. (1944). Studies on the chemical nature of the substance inducing transformation of pneumococcal types. Induction of transformation by a deoxyribonucleic acid fraction isoalted from Pneumococcus Type III. *J. Exp. Med.* **79**: 137.

Bailey-Serres, J. (1999). Selective translation of cytoplasmic mRNA in plants. *Trends Plant Science* **4**: 142–148.

Bakkali, A.T., Jaziri, M., Foriers, A., Vander Heyden, Y., Vanhaelen, M. and Homes, J. (1997). Lawsone accumulation in normal and transformed cultures of henna, *Lawsonia inermis*. *Plant Cell Tissue Organ Culture* **51**: 83–87.

Ball, J.M., Graham, D.Y., Opekun, A.R., Gilger, M.A., Guerrero, R.A. and Estes, M.K. (1999). Recombinant Norwalk virus-like particles given orally to volunteers: Phase I study. *Gastroenterology* **117**: 40–48.

Banerjee, S., Rahman, L., Uniyal, G.C. and Ahuja, P.S. (1998). Enhanced production of valepotriates by *Agrobacterium rhizogenes* induced hairy-root cultures of *Valeriana wallichii*. *Plant Science* **131**: 203–208.

Barthels, N., Van der Lee, F.M., Klap, J., Goddijn, O.J.M., Karimi, M., Puzio, P. Grundler, F.M.W., Ohl, S.A., Lindsey, K., Robertson, L., Robertson, W.M., Montagu, M. Van, Gheyesen, G. and Sijmons, P.C. (1997). Regulatory sequences of Arabidopsis drive reporter gene expression in nematode feeding structures. *Plant Cell* **9**: 2119–2134.

Baum, T.J., Hiatt, A., Parrott, W.A. and Hussey, R.S. (1996). Expression in tobacco of a functional monoclonal antibody specific to stylet secretions of the root-knot nematode. *Molec. Plant Microb. Interaction* **9**: 382–387.

Baumann, G., Raschke, E., Bevan, M. and Schoffl, F. (1987). Functional analysis of sequences required for transcriptional activation of a soybean heat shock gene in transgenic tobacco plants. *EMBO J.* **6**: 1161–1166.

Beachy, R.N., Fitchen, J.H. and Hein, M.B. (1996). Use of plant viruses for delivery of vaccine epitopes. *Ann. NY Acad. Sci.* **792**: 43–49.

Bechtold, N., Ellis, J. and Pelletier, G. (1993). *In planta Agrobacterium*-mediated gene transfer by infiltration of adult *Arabidopsis thaliana* plants. *CR Acad. Sci. Paris, Life Sciences* **316**: 1194–1199.

Bendahmane, M., Koo, M., Karrer, E. and Beachy, R.N. (1999). Display of epitopes on the surface of tobacco mosaic virus: Impact of charge and isoelectric point of the epitope on virus-host interaction. *J. Mol. Biol.* **290**: 9–20.

Berlin, J., Kuzovkina, I.N., Rugenhagen, C., Fecker, L., Commandeur, U. and Wray, V. (1992). Hairy root cultures of *Peganum harmala* II. Characterization of cell lines and effect of culture conditions on the accumulation of β-carboline alkaloids and serotonin. *Z. Naturforsch.* **47c**: 222–230.

Bhadra, R., Vani, S. and Shanks, J.V. (1993). Production of indole alkaloids by selected hairy root lines of *Catharanthus roseus*. *Biotech. Bioeng.* **41**: 581–592.

Bird, R.E., Hardman, K.D., Jacobson, J.W., Johnson, S., Kaufman, B.M., Lee, S.-M., Lee, T., Pope, S.H., Riordan, G.S. and Whitlow, M. (1988). Single-chain antigen-binding proteins. *Science* **242**: 423–426.

Bird, A.P. and Wolffe, A.P. (1999). Methylation-induced repression-belts, braces, and chromatin. *Cell* **99**: 451–454.

Bock, R. and Hagemann, R. (2000). Extranuclear inheritance: Plastid genetics: Manipulation of plastid genomes and biotechnological applications. *Progress in Botany* **61**: 76–90.

Bosch, D., Smal, J. and Krebbers, E. (1994). A trout growth hormone is expressed, correctly folded and partially glycosylated in the leaves but not the seeds of transgenic plants. *Transgenic Res.* **3**: 304–310.

Braun, A.C. (1969). Abnormal growth in plants. In: *Plant Physiology — A Treatise Vol. VB* (Steward, F.C., ed.) Academic Press, NY pp. 379–420.

Breiman, A. and Galun, E. (1990). Nuclear-mitochondrial interaction in angiosperms. *Plant Science* **71**: 3–19.

Brennan, F.R., Jones, T.D., Gilleland, L.B., Bellaby, T., Xu, F., North, P.C., Thompson, A., Staczek, J., Lin, T., Johnson, J.E., Hamilton, W.D.O. and Gilleland, H.E. Jr. (1999a). *Pseudomonas aeruginosa* outer membrane protein F epitopes are highly immunogenic in mice when expressed on a plant virus. *Microbiology* **145**: 211–220.

Brennan, F.R., Bellaby, T., Helliwell, S.M., Jones, T.D., Kamstrup, S., Dalsgaard, K., Flock, J.-I. and Hamilton, W.D.O. (1999b). Chimeric plant virus particles administered nasally or orally induce systemic and mucosal immune responses in mice. *J. Virol.* **73**: 930–938.

Brennan, F.R., Jones, T.D., Mongstaff, M., Chapman, S., Bellaby, T., Smith, H., Xu, F., Hamilton, W.D.O. and Flock, J.-I. (1999c). Immunogenicity of peptides derived from a fibronectin-binding protein of *S. aureus* expressed on two different plant viruses. *Vaccine* **17**: 1846–1857.

Brisson, N., Paszkowski, J., Penswick, J.R., Gronenborn, B., Potrykus, I. and Hohn, T. (1984). Expression of a bacterial gene in plants. *Nature* **310**: 511–514.

Browning, K.S. (1996). The plant translational apparatus. *Plant Mol. Biol.* **32**: 107–144.

Bruyns, A.-M., De Jaeger, G., Ne Neve, M., De Wilde, C., Van Montago, M. and Depicker, A. (1996). Bacterial and plant-produced scF_v proteins have similar antigen-binding properties. *FEBS Lett.* **386**: 5–10.

Caddick, M.X., Greenland, A.J., Jepson, I., Krause, K.-P., Qu, N., Riddell, K.V., Salter, M.G., Schuch, W., Sonnewald, U. and Tomsett, A.B. (1998). An ethanol inducible gene switch for plants used to manipulate carbon metabolism. *Nature Biotech.* **16**: 177–180.

Carrillo, C., Wigdorovitz, A., Oliveros, J.C., Zamorano, P.I., Sadir, A. M., Gomez, N., Salinas, J., Escribano, J.M. and Borca, M.V. (1998). Protective immune response to foot and mouth disease virus with VPI expressed in transgenic plants. *J. Virol.* **72**: 1688–1690.

Castanon, S., Marin, M.S., Martin-Alonso, J.M., Boga, J.A., Casais, R., Humara, J.M., Ordas, R.J. and Parra, F. 1999. Immunization with potato plants expressing VP60 protein protects against rabbit hemorrhagic disease virus. *J. Virol.* **73**: 4452–4455.

Chamberlain, D. and Stewart, C.N. Jr. (1999). Transgene escape and transplastomics. *Nature Biotech.* **17**: 330–331.

Chilton, M.-D, Tepfer, D., Petit, A., Chantal, D., Casse-Delbart, F. and Tempe, J. (1982). *Agrobacterium rhizogenes* inserts T-DNA into the genomes of the host plant root cells. *Nature* **295**: 432–434.

Chong, D.K.X., Roberts, W., Arakawa, T., Illes, K., Bagi, G., Slattery, C.W. and Langridge, W.H.R. (1997). Expression of the human milk protein β-casein in transgenic potato plants. *Transgenic Res.* **6**: 289–296.

Christen, P., Roberts, M.F., Phillipson, J.D. and Evans, W.C. (1989). High-yield production of tropane alkaloids by hairy-root cultures of a *Datura candida* hybrid. *Plant Cell Rep.* **8**: 75–77.

Christou, P. (1996). Transformation technology. *Trends Plant Science* **1**: 423–431.

Clough, S.J. and Bent, A.F. (1998). Floral dip: A simplified method for *Agrobacterium*-mediated transformation of *Arabidopsis thaliana*. *Plant J.* **16**: 735–743.

Conrad, U. and Fiedler, U. (1998). Compartment-specific accumulation of recombinant immunoglobulins in plant cells: An essential tool for antibody production and immunomodulation of physiological functions and pathogen activity. *Plant Mol. Biol.* **38**: 101–109.

Conrad, U., Fiedler, U., Artsaenko, O. and Phillips, J. (1998a). Single-chain F_v antibodies expressed in plants. In: *Methods in Biotechnology Vol 3: Production and Isolation of Clinically Useful Compounds.* (Cunningham, C. and Porter, A.Y.R., eds.) Humana Press, Totowa, NJ, pp. 103–127.

Conrad, U., Fiedler, U., Artsaenko, O. and Phillips, J. (1998b). High-level and stable accumulation of single-chain F_v antibodies in plant storage organs. *J. Plant Physiol.* **154**: 708–711.

Cramer, C.L., Boothe, J.G. and Oishi, K.K. (1999). Transgenic plants for therapeutic proteins: linking upstream and downstream strategies. In: *Plant Biotechnology — New Products and Applications* (Hammon, J., McGravey, P. and Yusibov, V., eds.) Springer-Verlag, Heidelberg, pp. 95–118.

Cramer, C.L., Weissenborn, D.L., Oishi, K.K., Grabau, E.A., Bennett, S., Ponce, E., Grabowski, G.A. and Radin, D.N. (1996). Bio-production of human enzymes in transgenic tobacco. *Ann. NY Acad. Sci.* **792**: 62–71.

Dalsgaard, K., Uttenthal, A., Jones, T.D., Xum, F., Merryweather, A., Hamilton, W.D.O., Langeveld, J.P.M., Boshuizen, R.S., Kamstrup, S., Lomonossoff, G.P., Porta, C., Vela, C., Casal, J.I., Meloen, R.H. and Rodgers, P.B. (1997). Plant-derived vaccine protects target animals against a viral disease. *Nature Biotech.* **15**: 248–252.

Daniell, H. (1999). Environmentally friendly approaches to genetic engineering. *In Vitro Cell Devel. Biol. Plant* **35**: 361–368.

Dasgupta, P. and Mukherjee, R. (2000). Lipophilization of somatostatin analog RC-160 with long chain fatty acid improves its anti-proliferative and antiogenic activity *in vitro. Br. J. Pharmacol.* **129**: 101–109.

De Block, M., Schell, J. and Van Montagu, M. (1985). Chloroplast transformation by *Agrobacterium tumefaciens*. *EMBO J.* **4**: 1367–1372.

De Jaeger, G., Buys, E., Eeckhout, D., Bruyns, A.-M., De Neve, M., De Wilde, C, Gerats, T., Van Montagu, M., Fischer, R. and Depicker, A. (1997). Use of phage display for isolation and characterization of single-chain variable fragments against dihydroflavonol 4-reductase from *Petunia hybrida*. *FEBS Lett.* **403**: 116–122.

Deno, H., Yamagata, H., Emoto, T., Yoshioka, T., Yamada, Y. and Fujita, Y. (1987). Scopolamine production by root cultures of *Duboisia myoporoides*. II. Establishment of a hairy-root culture by infection with *Agrobacterium rhizogenes*. *J. Plant Physiol.* **131**: 315–323.

De Neve, M., De Buck, S., De Wilde, C., Van Houdt, H., Strobbe, I., Jacobs, A., Van Montagu, M. and Depicker, A. (1999). Gene silencing results in instability of antibody production in transgenic plants. *Mol. Gen. Genet.* **260**: 582–592.

Denecke, J., Botterman, J. and Deblaere, R. (1990). Protein secretion in plant cells can occur via a default pathway. *Plant Cell* **2**: 51–59.

De Santis-Macioszek, G., Kofer, W., Bock, A., Schoch, S., Maier, R.M., Wanner, G., Rudiger, W., Koop, H.-U. and Herrmann, R.G. (1999). Targeted disruption of the plastid RNA polymerase genes *rpo*A, B and C1: Molecular biology, biochemistry and ultrastructure. *Plant J.* **18**: 477–489.

De Wilde, C., De Neve, M., De Rycke, R., Bruyns, A.-M., De Jaeger, G., Van Montagu, M., Depicker, A. and Engler, G. (1996). Intact antigen-binding MAK33 antibody and F_{ab} fragment accumulate in intercellular spaces of *Arabidopsis thaliana*. *Plant Science* **114**: 233–241.

De Zoeten, G.A., Penswick, J.R., Horisberger, M.A., Ahl, P., Schultze, M. and Hohn, T. (1989). The expression, localization, and effect of a human interferon in plants. *Virology* **172**: 213–222.

Domansky, N., Ehsani, P., Salmanian, A.H. and Medvedeva, T. (1995). Organ specific expression of hepatitis B surface antigen in potato. *Biotech. Lett.* **17**: 863–866.

Donson, H.J., Kearney, C.M., Hilf, M.E. and Dawson, W.O. (1991). Systemic expression of a bacterial gene by a tobacco mosaic virus-based vector. *Proc. Natl. Acad. Sci. USA* **88**: 7204–7208.

Durand-Tardif, M., Broglie, R., Slightom, J. and Tepfer, D. (1985). Structure and expression of *ri* T-DNA from *Agrobacterium rhizogenes* in *Nicotiana tabacum*. *J. Mol. Biol.* **186**: 557–564.

Düring, K., Hippe, S., Kreuzaker, F. and Schell, J. 1990. Synthesis and self-assembly of a functional monoclonal antibody in transgenic *Nicotiana tabacum*. *Plant Mol. Biol.* **15**: 281–293.

Durrani, Z., McInerney, T.L., McLain, L., Jones, T., Bellaby, T., Brennan, Fr. And Dimmock, N.J. 1998. Intranasal immunization with a plant virus expressing a peptide from HIV-1 gp41 stimulates better mucosal and systemic HIV-1-specific IgA and IgG than oral immunization. *J. Immun. Meth.* **220**: 93–103.

Endo, T., Goodbody, A. and Misawa, M. (1987). Alkaloid production in root and shoot cultures of *Catharanthus roseus*. *Planta Medica* **53**: 479–482.

Ehsani, P., Khabiri, A. and Domansky, N.N. (1997). Polypeptides of hepatitis B surface antigen produced in transgenic potato. *Gene* **190**: 107–111.

Fecker, L.F., Rügenhagen, C. and Berlin, J. (1993). Increased production of cadaverine and anabasine in hairy root cultures of *Nicotiana tabacum* expressing a bacterial lysine decarboxylase gene. *Plant Mol. Biol.* **23**: 11–21.

Feldmann, K.A. (1991). T-DNA insertion mutagenesis in *Arabidopsis:* Mutational spectrum. *Plant J.* **1**: 71–82.

Fenner, F., Henderson, D.A., Arita, L., Jezek, Z. and Ladnyi, I.D. (1988). Smallpox and its eradication. World Health Organization Publications, Geneva.

Fettig, S. and Hess, D. (1999). Expression of a chimeric stilbene synthase gene in transgenic wheat lines. *Transgenic Res.* **8**: 179–189.

Fiedler, U. and Conrad, U. (1995). High-level prodcution and long-term storage of engineered antibiotics in trasngenic tobacco seeds. *Bio/Technology* **13**: 1090–1093.

Finer, J.J., Finer, K.R. and Ponceppa, T. (1999). Particle-bombardment-mediated transformation. In: *Plant Biotechnology — New Products and Applications* (Hammon, J., McGravey, P. and Yusibov, V., eds.) Springer-Verlag, Heidelberg, pp. 59–80.

Firek, S., Draper, J., Owen, M.R.L., Gandecha, A., Cockburn, B. and Whitelam, G.C. (1993). Secretion of a functional single-chain F_v protein in transgenic tobacco plants and cell suspension cultures. *Plant Mol. Biol.* **23**: 861–870.

Fischer, R., Drossard, J., Liao, Y.-C. and Schillberg, S. (1998). Characterization and applications of plant-derived recombinant antibodies. In: *Methods in Biotechnology. Vol. 3: Production and Isolation of Clinically Useful Compounds* (Cunningham, C. and Porter, A.J.R., eds.) Humana Press, NJ, pp. 129–142.

Fischer, R., Drossard, J., Commandeur, U., Schillberg, S. and Emans, N. (1999a). Toward molecular farming in the future: moving from diagnostic protein and antibody production in microbes to plants. *Biotechnol. Appl. Biochem.* **30**: 101–108.

Fischer, R., Emans, N., Schuster, F., Hellwig, S. and Drossard, J. (1999d). Toward molecular farming in the future: Using plant cell suspension cultures as bioreactors. *Biotechnol. Appl. Biochem.* **30**: 109–112.

Fischer, R., Liao, Y.-C., Hoffmann, K., Schillberg, S. and Emans, N. (1999b). Molecular farming of recombinant antibodies in plants. *Biol. Chem.* **380**: 825–839.

Fischer, R., Schumann, D., Zimmermann, S., Drossard, J., Sack, M. and Schillberg, S. (1999c). Expression and characterization of bispecific single chain F_v fragments produced in transgenic plants. *Eur. J. Biochem.* **262**: 810–816.

Fitchen, J., Beachy, R.N. and Hein, M.B. (1995). Plant virus expressing hybrid coat protein with added murine epitope elicits auto-antibody response. *Vaccine* **13**: 1015–1057.

Flaming, D.O., Richardson, J.H., Tulis, J.L. and Veseley, D.O. (1995). *Laboratory Safety, Principles and Practices*, ASM Press, Washington DC.

Flores, H.E., Hoy, M.W. and Pickard, J.J. (1987). Secondary metabolites from root cultures. *Trends Biotech.* **5**: 64–69.

Furze, J.M., Hamill, J.D., Parr, A.J., Robins, R.J. and Rhodes, M.J.C. (1987). Variations in morphology and nicotine alkaloid accumulation in protoplast-derived hairy root cultures of *Nicotiana rustica*. *J. Plant Physiol.* **131**: 237–246.

Futter, J. and Hohn, T. (1996). Translation in plants — Rules and exceptions. *Plant Mol. Biol.* **32**: 156–189.

Futterer, J. (1995). Expression signals and vectors. In: *Gene Transfer to Plants* (Potrykus, I and Spangenberg, G., eds) Springer-Verlag, Berlin, pp. 311–324.

Galili, G., Sengupta-Gopalan, C. and Ceriotti, A. (1998). The endoplasmic reticulum of plant cells and its role in protein maturation and biogenesis of oil bodies. *Plant Mol. Biol.* **38**: 1–29.

Galili, S., Fromm, H., Aviv, D., Edelman, M. and Galun, E. (1989). Ribosomal protein S12 as a site for streptomycin resistance in *Nicotiana* chloroplasts. *Mol. Gen. Genet.* **218**: 289–292.

Gallie, D.R. (1996). Translational control of cellular and viral mRNAs. *Plant Mol. Biol.* **32**: 145–158.

Galun, E. (1981). Plant protoplasts as physiological tools. *Ann. Rev. Plant Physiol.* **32**: 237–266.

Galun, E. and Aviv, D. (1986). Organelle transfer. *Meth. Enzymol.* **118**: 595–611.

Galun, E. and Breiman, A. (1997). *Transgenic Plants.* Imperial College Press, London, 376 pp.

Gandecha, A., Owen, M.R.L., Cockburn, W. and Whitelam, G.C. (1994). Antigen detection using recombinant, bifunctional single-chain F_v fusion proteins synthesised in *Escherichia coli. Protein Exp. Purific.* **5**: 385–390.

Gatz, C. (1997). Chemical control of gene expression. *Ann. Rev. Plant Physiol. Plant Mol. Biol.* **48**: 89–108.

Gatz, C. and Lenk, I. (1998). Promoters that respond to chemical inducers. *Trends Plant Science* **3**: 352–358.

Gehm, B.D., McAndrews, J.M., Chien, P.-Y. and Jameson, J.L. (1997). Resveratrol, a polyphenolic compound found in grapes and wine, is an agonist for the estrogen receptor. *Proc. Natl. Acad. Sci. USA* **94**: 14138–14143.

Gindullis, F. and Meier, I. (1999). Matrix attachment region binding protein MFP1 is localized in discrete domains at the nuclear envelope. *Plant Cell* **11**: 1117–1128.

Givol, D. (1991). The minimal antigen-binding fragment of antibodies — F_v fragment. *Mol. Immunol.* **28**: 1379–1386.

Golds, T., Maliga, P. and Koop, H.-U. (1993). Stable plastid transformation in PEG-treated protoplasts of *Nicotiana tabacum. Bio/Technology* **11**: 95–97.

Goldschmidt, R. (1955). *Theoretic Genetics*, University of California Press, Berkeley.

Gomez, N., Carrillo, C., Salinas, J., Parra, F., Borca, M.V. and Escribano, J.M. (1998). Expression of immunogenic glycoprotein S poypeptides from transmissible gastroenteritis coronavirus in transgenic plants. *Virology* **249**: 352–358.

Gosch-Wackerle, G., Avivi L. and Galun, E. (1979). Induction, culture and differentiation of callus from immature rachises, seeds and embryos of *Triticum. Z. Pflanzenphysiol.* **91**: 267–278.

Gräuicher, F., Christen, P. and Kapetanidis, T. (1992). High-yield production of valepotriates by hairy-root cultures of *Valeriana officinalis* var. *sambucifolia. Plant Cell Rep.* **11**: 339–342.

Gray, M.W. (1993). Origin and evolution of organelle genomes. *Curr. Opin. Genet. Dev.* **3**: 884–890.

Gressel, J. (1999). Tandem constructs; preventing the rise of super-weeds. *Trends Biotech.* **17**: 361–366.

Gressel, J. and Rotteveel, T. (2000). Genetic and ecological risks from biotechnologically-derived herbicide-resistant crops: Decision trees for risk assessment. *Plant Breeding Rev.* **18**: 251–303.

Gutierrez, R.A., MacIntosh, G.C. and Green, P.J. (1999). Current perspectives on mRNA stability in plants: Multiple levels and mechanisms of control. *Trends Plant Science* **4**: 429–438.

Hadi, M.Z., McMullen, M.D. and Finer, J.J. (1996). Transformation of 12 different plasmids into soybean via particle bombardment. *Plant Cell Rep.* **15**: 500–505.

Hamamoto, H., Sugiyama, Y., Nakagawa, N., Hishida, E., Matsunaga, Y., Takemoto, S., Watanabe, H. and Okada, Y. (1993). A new tobacco mosaic virus vector and its use for the systemic production of angiotensin-I-converting enzyme inhibitor in transgenic tobacco and tomato. *Bio/Technology* **11**: 930–932.

Hamill, J.D., Parr, A.J., Robins, R.J. and Rhodes, M.J.C. (1986). Secondary product formation by cultures of *Beta vulgaris* and

Nicotiana rustica transformed with *Agrobacterium rhizogenes*. *Plant Cell Rep*. **5**: 111–114.

Hamill, J.D., Robins, R.J. and Rhodes, M.J.C. (1989). Alkaloid production by transformed root cultures of *Cinchona ledgeriana*. *Planta Medica* **55**: 354–357.

Hammond, J. (1999). Overview: The many uses and applications of transgenic plants. In: *Plant Biotechnology-New Products and Applications* (Hammon, J., McGarvey, P. and Yusibov, V., eds.) Springer-Verlag, Heidelberg, pp. 1–19.

Hammond, J., McGarvey, P. and Yusibov, V. (eds.) (1999). *Plant Biotechnology: New Products and Applications*, Springer-Verlag, Heidelberg.

Han, K.-H., Caiping, M.A. and Strauss, S.H. (1997). Matrix attachment regions (MARs) enhance transformation frequency and transgene expression in poplar. *Transgenic Res*. **6**: 415–420.

Hansen, G. and Chilton, M.D., (1999). Lessons in gene transfer to plant by a gifted microbe. In: *Plant Biotechnology — New Products and Applications* (Hammon, J., McGravey, P. and Yusibov, V., eds.) Springer-Verlag, Heidelberg, pp. 21–57.

Hansen, G. and Wright, M.S. (1999). Recent advances in the transformation of plants. *Trends Plant Science* **4**: 226–231.

Hansen, G., Shillito, R.D. and Chilton, M.-D. (1997). T-strand integration in maize protoplasts after co-delivery of a T-DNA substrate and virulence genes. *Proc. Natl. acad. Sci. USA* **94**: 11726–11730.

Haq, T.A., Mason, H.S., Clements, J.D. and Arntzen, C.J. (1995). Oral immunization with a recombinant bacterial antigen produced in transgenic plants. *Science* **268**: 714–716.

Harlow, E. and Lane, D (1988). *Antibodies: A Laboratory Manual*. Cold Spring Harbor Laboratory Press, Cold Spring Harbor, NY.

Hashimoto, T., Matsuda, J. and Yamada, Y. (1993). Two-step epoxidation of hyoscyamine to scopolamine is catalysed by bifunctional hyoscyamine-β6-hydroxylase. *FEBS Lett.* **329**: 35–39.

Haynes, J.R., Cunningham, J., von Seefried, A., Lennick, M., Garvin, R.T. and Shen, S.H. (1986). Engineered, candidate polio vaccine employing the self-assemblying proprties of the tobacco mosaic virus coat protein. *Bio/Technology* **4**: 637–641.

Herman, E.M. and Larkins, B.A. (1999). Protein storage bodies and vacuoles. *Plant Cell* **11**: 601–613.

Herrmann, R.G. (1997). Eukaryotism towards a new interpretation. In: *Eukaryotism and Symbiosis* (Schenk, H.E.A., Herrmann, R.G., Jeon, K.W., Müller, N.E. and Schwemmler, W., eds.) Springer-Verlag, New York, pp. 206–213.

Hiatt, A. (1990). Antibodies produced in plants. *Nature* **344**: 469–470.

Hiatt, A., Cafferkey, R. and Bowdish, K. (1989). Production of antibodies in transgenic plants. *Nature* **342**: 76–78.

Higo, K.-I., Siato, Y. and Higo, H. (1993). Expression of a chemically synthesized gene for human epidermal growth factor under the control of cauliflower mosaic virus 35S promoter in transgenic tobacco. *Biotech. Biochem.* **57**: 1477–1481.

Hillman, M.R. (1999). Personal historical chronicle of six decades of basic and applied research in virology, immunology and vaccinology. *Immunol. Rev.* **170**: 7–27.

Hochman, J., Gavish, M., Inbar, D. and Givol, D. (1976). Folding and interaction of subunits at the antibody combining site. *Biochem.* **15**: 2706–2710.

Hood, E.E., Kusnadi, A., Nikolov, Z. and Howard, J.A. (1999). Molecular farming of industrial proteins from transgenic maize. In: *Chemicals via Higher Plant Bioengineering* (Shahidi *et al.*, eds.) Plenum Pub., NY, pp. 127–147.

Hood, E.E., Witcher, D.R., Maddock, S., Meyer, T., Baszczynski, C., Bailey, M., Flynn, P., Register, J., Marshall, L., Bond, D., Kulisek, E., Kusnadi, A., Evangelista, R., Nikolov, Z., Wooge, C., Mehigh, R.J., Herman, R., Kappel, W.K., Ritland, D., Li, C.-P. and Howard, J.A. (1997). Commercial production of avidin from transgenic maize: characterization of transformant production processing extraction and purification. *Mol. Breeding* **3**: 291–306.

Hu, Z.B. and Alfermann, A.W. (1993). Diterpenoid production in hairy root cultures of *Salvia miltiorrhiza*. *Phytochem.* **32**: 699–703.

Hull, R. (1978). The possible use of plant viral DNAs in genetic manipulation of plants. *Trends Biochem. Sci.* **3**: 254–256.

Huston, J.S., Levinson, D., Mudgett-Hunter, M., Tai, M.-S., Novotny, J., Margolies, M.N., Ridge, R.J., Burccoleri, R.E., Haber, E., Crea, R. and Oppermann, H. (1988). Protein engineering of antibody binding sites: Recovery of specific activity in an anti-digoxin single-chain Fv analogue produced in *Escherichia coli*. *Proc. Natl. Acad. Sci. USA* **85**: 5879–5883.

Inbar, D., Hochman, J. and Givol, D. (1972). Localization of antibody-combining sites within the variable portions of heavy and light chains. *Proc. Natl. Acad. Sci. USA* **69**: 2659–2662.

Ingelbrecht, L.W., Herman, L.M.F., Dekeyser, R.A., van Montagu, M.C. and Depicker, A.G. (1996). Different 3′ end regions strongly influence the level of gene expression in plant cells. *Plant Cell* **1**: 579–589.

Janeway, C.A. Jr. and Travers, P. (1994). *Immunobiology: The Immune System in Health and Disease*. Current Biology Ltd., London.

Jang, M., Cai, L., Udeani, G.O., Slowing, K.V., Thomas, C.F., Beecher, C.W.W., Fong, H.H.S., Farnsworth, N.R., Kinghorn, A.D., Mehta, R.G., Moon, R.C. and Pezzuto, J.M. Cancer chemopreventive activity of resveratrol, a natural product derived from grapes. *Science* **275**: 218–220.

Jaziri, M., Legros, M., Homes, J. and Vanhaelen, M. (1988). Tropine alkaloids production by hairy root cultures of *Datura stramonium* and *Hyoscyamus niger*. *Phytochem.* **27**: 419–420.

Jefferson, R.A., Burgess, S.M. and Hirsch, D. (1986). β-glucuronidase from *Escherichia coli* as a gene-fusion marker. *Proc. Natl. Acad. Sci. USA* **83**: 8447–8451.

Joelson, T., Akerblom, L., Oxelfels, P., Strandberg, B., Tomenius, K. and Morris, T.J. (1997). Presentation of a foreign peptide on the surface of tomato bushy stunt virus. *J. Gen. Virol.* **78**: 1213–1217.

Joersbo, M., Donaldson, I., Kreiberg, J., Guldager Petersen, S., Brunstedt, J and Okkels, F.T. (1998). Analysis of mannose selection used for transformation of sugarbeet. *Mol. Breeding* **4**: 111–117.

Johnson, J., Lin, T. and Lomonossoff, G. (1997). Presentation of heterologous peptides on plant viruses. *Annu. Rev. Phytopathol.* **35**: 67–86.

Jones, P.T., Dear, P.H., Foote, J., Neuberger, M.S. and Winter, G. (1986). Replacing the complementary-determining regions in a human antibody with those from a mouse. *Nature* **321**: 522–525.

Jouhikainen, K., Lindgren , L., Jokelainen, T., Hiltunen, R., Teeri, T.H. and Oksman-Caldentey, K.-M. (1999). Enhancement of scopolamine production in *Hyoscyamus muticus* L. Hairy-root cultures by genetic engineering. *Planta* **208**: 545–551.

Jung, G. and Tepfer, D. (1987). Use of genetic transformation by the Ri T-DNA of *Agrobacterium rhizogenes* to stimulate biomass and tropane alkaloid production in *Atropa belladonna* and *Calystegia sepium* roots grown *in vitro*. *Plant Science* **50**: 145–151.

Kaluza, B., Betzl, G., Shao, H., Diamantstein, T. and Weidle, U.H. (1992). A general method for chimerization of monoclonal antibodies by inverse polymerase chain reaction which conserves authentic N-terminal sequences. *Gene* **122**: 321–328.

Kamada, H., Okamura, N., Satake, M., Harada, H. and Shimomura, K. (1986). Alkaloid production by hairy root cultures in *Atropa belladonna*. *Plant Cell Rep.* **5**: 239–242.

Kapusta, J., Modelska, A., Figlerowicz, M., Pniewski, T., Letellier, M., Lisowa, O., Yusibov, V., Koprowski, H., Plucienniczak, A. and Legocki, A.B. (1999). A plant-derived edible vaccine against hepatitis B virus. *FASEB J.* **13**: 1796–1799.

Katagiri, F. and Chua, N.H. (1992). Plant transcription factors: Present knowledge and future challenges. *Trends Genet.* **8**: 22–27.

Khan, M.S. and Maliga, P. (1999). Fluorescent antibiotic resistance marker for tracking plastid transformation in higher plants. *Nature Biotech.* **17**: 910–915.

Kittipongpatana, N., Hock, R.S. and Porter, J.R. (1998). Prodcution of solasodine by hairy root, callus, and cell suspension cultures of *Solanum aviculare* Forst. *Plant Cell Tissue Organ Culture* **52**: 133–143.

Knoepfler, P.S. and Eisenman, R.N. (1999). Sin meets NuRD and other tails of repression. *Cell* **99**: 447–450.

Kohler, G. and Milstein, C. (1975). Continuous culture of fused cells seceting antibody of predefined specificity. *Nature* **256**: 495–497.

Koo, M., Bendahmane, M., Lettieri, G.A., Paoletti, A.D., Lane, T.E., Fitcher, J.H., Buchmeier, M.J. and Beachy, R.N. (1999). Protective immunity against murine hepatitis virus (MHV) induced by intranasal or subcutaneus administration of hybrids of tomato mosaic virus that carries an MHV epitope. *Proc. Natl. Acad. Sci. USA* **96**: 7774–7779.

Koop, H.-U. and Kofer, W. (1995). Plastid transformation by polyethylene glycol treatment of protoplasts and regeneration of transplatomic tobacco plants. In: *Gene Transfer to Plants* (Potrykus, I. and Spangenberg, G., eds.) Springer-Verlag, Berlin, pp. 75–82.

Koop, H.-U., Steinmuller, K., Wagner, H., Rosser, Eibl, C. and Sacher, L. (1996). Integration of foreign sequences into the tobacco plastome via PEG-mediated protoplast transformation. *Planta* **199**: 601–604.

Koziel, M.G., Carozzi, N.B. and Desai, N. (1996). Optimizing expression of transgenes with emphasis on post transcriptional events. *Plant Mol. Biol.* **32**: 393–405.

Kudla, J., Hayes, R. and Gruissem, W. (1996). Polyadenylation accelerates degradation of chloroplast mRNA. *EMBO J.* **15**: 7137–7146.

Kumagai, M.H., Turpen, T.H., Weinzettl, N., Della-Cioppa, G., Turpen, A.M., Donson, J., Hilf, M.E., Grantham, G.L., Dawson, W.O., Chow, T.P., Piatak, M. Jr. and Grill, L.K. (1993). Rapid, high-level expression of biologically active α–trichosanthin in transfected plants by an RNA viral vector. *Proc. Natl. Acad. Sci. USA* **90**: 427–430.

Laherty, C.D., Billin, A.N., Lavinsky, R.M., Yochum, G.S., Bush, A.C., Sun, J.-M., Mullen, T.-M., Davie, J.R., Rose, D.W., Glass, C.K., Rosenfeld, M.G., Ayer, D.E. and Eisenman, R.B. (1998). SAP30, a component of the mSin3 corepressor complex involved in N-CoR-mediated repression by specific transcription factors. *Mol. Cell.* **2**: 33–42.

Lebeurier, G., Hirth, L., Hohn, B. and Hohn, T. (1982). *In vitro* recombination of cauliflower mosaic virus DNA. *Proc. Natl. Acad. Sci. USA* **79**: 2932–2936.

Liu, C.Z., Wang, Y.C., Ouyang, F., Ye, H.C. and Li, G.F. (1998). Production of artemisinin by hairy root cultures of *Artemisia annua* L in bioreactor. *Biotechnol. Lett.* **20**: 265–268.

Liu, M.A. (1998). Vaccine development. *Nature Medicine* **4**: 515–519.

Lodhi, A.H., Bongaerts, R.J.M., Verpoorte, R., Coomber, S.A. and Charlwood, B.V. (1996). Expression of bacterial isochorismate synthase (EC 5.4.99.6) in transgenic root cultures of *Rubia peregrina. Plant Cell Rep.* **16**: 54–57.

Lomonossoff, G.P. and Hamilton, W.D.O. (1999). Cowpea mosaic virus-based vaccines. In: *Plant Biotechnology — New Products and Applications* (Hammon, J., McGravey, P. and Yusibov, V., eds.), Springer-Verlag, Heidelberg, pp. 177–189.

Lomonossoff, G. and Johnson, J.E. (1995). Eukaryotic viral expression systems for polypeptides. *Virology* **6**: 257–267.

Ma, J.K.-C. and Hein, M.B. (1995). Plant antibodies for immunotherapy. *Plant Physiol.* **109**: 341–346.

Ma, J.K.-C. and Hein, M.B. (1996). Antibody production and engineering in plants. *Ann. NY Acad. Sci.* **792**: 72–81.

Ma, J.K.-C., Hiatt, A., Hein, M., Vine, N.D., Wang, F., Stabila, P., van Dolleweerd, C., Mostov, K. and Lehner, T. (1995). Generation and assembly of secretory antibodies in plants. *Science* **268**: 716–719.

Ma, J.K.-C., Hikmat, B.Y., Wycoff, K., Vine, N.D., Chargelegue, D., Yu, L., Hein, M.B. and Lehner, T. (1998). Characterization of a recombinant plant monoclonal secretory antibody and preventive immunotherapy in humans. *Nature Medicine* **4**: 601–606.

Ma, J.K.-C., Lehner, T., Stabila, P., Fux, C.I. and Hiatt, A. (1994). Assembly of monoclonal antibodies with IgG1 and IgA heavy chain domains in transgenic tobacco plants. *Eur. J. Immunol.* **24**: 131–138.

Ma, J.K.-C. and Vine, N.D. (1999). Plant expression systems for the production of vaccines. *Curr. Topics Microbiol. Immunol.* **236**: 275–292.

Ma, S.-W., Zhao, D.-L., Yin, Z.-Q., Mukherjee, R., Singh, B., Qin, H.-Y., Stiller, C.R. and Jevnikar, A.M. (1997). Transgenic plants expressing autoantigens fed to mice to induce oral immune tolerance. *Nature Medicine* **3**: 793–796.

Mackenzie, S. and McIntosh, L. (1999). Higher plant mitochondria. *Plant Cell* **11**: 571–585.

Magnuson, N.S., Linzmaier, P.M., Reeves, R., Gynheung, A., HayGlass, K. and Lee, J.M. (1998). Secretion of biologically active human interleukin-2 and interleukin-4 from genetically modified tobacco cells in suspension culture. *Prot. Exp. Purific.* **13**: 45–52.

Mahler, J. (1999). New issues and future legislation on biosafety. *J. Biotech.* **68**: 179–183.

Maldonado, E., Hampsey, M. and Reinberg, D. (1999). Repression: Targeting the heart of the matter. *Cell* **99**: 455–458.

Maliga, P., Moll, B. and Svab, Z. (1990). Towards manipulation of plastid genes in higher plants. In: *Perspectives in Biochemical and Genetic Regualtion of Photosynthesis* (Zeitlich, I., ed.) Alan R. Liss, Inc., pp. 133–143.

Mano, Y., Ohkawa, H. and Yamada, Y. (1989). Production of tropane alkaloids by hairy root cultures of *Duboisia leichhardtii* transformed by *Agrobacterium rhizogenes*. *Plant Science* **59**: 191–201.

Marden, M.C., Dieryck, W., Pagnier, J., Pyart, C., Gruber, V., Bournat, P., Olagnier, B, Theisen, M. and Merot, B. (1998). Transgenic plants as an alternate source of human hemoglobin. *Proceedings of Transgenic Production of Human Therapeutics*, Waltham, MA, USA, pp. 1/4–4/4.

Margulis, L. (1981). *Symbiosis in Cell Evolution*. W.H. Freeman and Co., San Francisco, 313 pp.

Mason, H.S., Ball, J.M., Shi, J.-J., Jiang, X., Estes, M.K. and Arntzen, C.J. (1996). Expression of Norwalk virus capsid protein in transgenic tobacco and potato and its oral immunogenicity in mice. *Proc. Natl. Acad. Sci. USA* **93**: 5335–5340.

Mason, H.S., Haq, T.A., Clements, J.D. and Arntzen, C.J. (1998). Edible vaccine protects mice against *Escherichia coli* heat-labile enterotoxin (LT): Potatoes expressing a synthetic LT-B gene. *Vaccine* **16**: 1336–1343.

Mason, H.S., Lam, D.M.-K. and Arntzen, C.J. (1992). Expression of hepatitis B surface antigen in transgenic plants. *Proc. Natl. Acad. Sci. USA* **89**: 11745–11749.

Matsumoto, S., Ikura, K., Ueda, M. and Sasaki, R. (1995). Characterization of a human glycoprotein (erythropoietin) produced in cultured tobacco cells. *Plant Mol. Biol.* **27**: 1163–1172.

McCormick, A.A., Kumagai, M.H., Hanley, K., Turpen, T.H., Hakim, I., Grill, L.K., Tuse, D., Levy, S. and Levy, R. (1999). Rapid production of specific vaccines for lymphoma by expression of the tumor-derived single-chain F_v epitopes in tobacco plants. *Proc. Natl. Acad. Sci. USA* **96**: 703–708.

McGarvey, P.B., Hammond, J., Dienelt, M.-M., Hooper, D.C., Fu, F.Z., Dietzschold, B., Koprowski, H. and Michaels, F.H. (1995). Expression of the rabies virus glycoprotein in transgenic tomatoes. *Bio/Technology* **13**: 1484–1487.

McInerney, T.L., Brennan, F.R., Jones, T.D. and Dimmock, N.J. (1999). Analysis of the ability of five adjuvants to enhance immune responses to a chimeric plant virus displaying an HIV-1 peptide. *Vaccine* **17**: 1359–1368.

McLain, L., Durrani, Z., Wisniewski, L.A., Porta, C., Lomonossoff, G.P. and Dimmock, N.J. (1996). Stimulation of neutralizing antibodies to human immunodeficiency virus type 1 in three stains of mice immunized with a 22 amino acid peptide of gp41 expressed on the surface of a plant virus. *Vaccine* **14**: 799–810.

McLain, L., Porta, C., Lomonossoff, G.P., Durrani, Z. and Dimmock, N.J. (1995). Human imminodeficiency virus type 1-neutralizing antibodies raised to a glycoprotein 41 peptide expresssed on the surface of a plant virus. *AIDS Res. Human Retroviruses* **11**: 327–334.

Mekalanos, J.J. and Sadaff, J.C. (1994). Cholera vaccines: Fighting an ancient scourge. *Science* **265**: 1387–1389.

Mereschkowsky, K. (1910). Theorie der zwei plasmaarten als Grundlage der Symbiogenesis einer neuer lehre von der eststahung der Organismen. *Biologisches Centralblatt* **30**: 352–367.

Meshi, T. and Iwabuchi, M. (1995). Plant transcription factors. *Plant Cell Physiol.* **36**: 1405–1420.

Mett, V.L., Lochhead, L.P. and Reynolds, P.H.S. (1993). Copper controllable gene expression system for whole plants. *Proc. Natl. Acad. Sci. USA* **90**: 4567–4571.

Mett, V.L., Podivinsky, E., Tennant, A.M., Lochhead, L.P., Jones, W.T. and Reynolds, P.H.S. (1996). A system for tissue-specific copper controllable gene expression in transgenic plants: Nodule-specific antisense of asparate aminotransferase-P2. *Transgenic Res.* **5**: 105–113.

Mett, V.L. and Reynolds, P.H.S. (1999). Tissue specific copper-controllable gene expression in plants. In: *Inducible Gene Expression in Plants* (Reynolds, P.H.S., ed.) CABI Publishing, Wallingford (UK), pp. 61–81.

Minivelle-Sebastia, L. and Keller, W. (1999). mRNA polyadenylation and its coupling to other RNA processing reactions and to transcription. *Curr. Opin. Cell Biol.* **11**: 352–357.

Mitra, A. and Zhang, Z. (1994). Expression of a human lactoferrin cDNA in tobacco cells produces antibacterial protein(s). *Plant Physiol.* **106**: 977–981.

Mittelstem-Scheid, O., Afsar, K. and Paszkowski, J. (1998). Release of epigenic gene silencing by transacting mutations in *Arabidopsis*. *Proc. Natl. Acad. Sci. USA* **95**: 632–637.

Moffat, A.S. (1995). Exploring transgenic plants as a new vaccine source. *Science* **268**: 658–660.

Moore, D.J. and Mollenhauer, H.H. (1971). The endomenmbrane concept: A functional integration of endoplasmic reticulum and

Golgi apparatus. In: *Dynamic Aspects of Plant Ultrastructure.* (Robaards, A.W., ed.) McGraw-Hill, London, pp. 84–137.

Mor, T., Gomez-Lim, M.A. and Palmer, K.E. (1998). Perspective edible vaccines — A concept coming of age. *Trends Microbiol.* **6**: 449–453.

Morgan, J.A. and Shanks, J.V. (2000). Determination of metabolic rate-limitations by precursor feeding in *Catharanthus roseus* hairy-root cultures. *J. Biotech.* **79**: 137–146.

Mukundan, Y. and Hjortso, M.A. (1991). Growth and thiophene accumulation by hairy root cultures of *Tagetes patula* in media of varying initial pH. *Plant Cell Rep.* **9**: 627–630.

Muranaka, T., Ohkawa, H. and Yamada, Y. (1992). Scopolamine release into media by *Duboisia leichhardtii* hairy root clones. *Appl. Microbiol Biotechnol.* **37**: 554–559.

Murashige, T. and Skoog, F. (1962). A revised medium for rapid growth and bioassays with tobacco tissue culture. *Physiol. Plant.* **15**: 493–497.

Nehra, N.S., Chibbar, R.N., Leung, N., Caswell, K., Mallard, C., Steinhauer, L., Baga, M. and Kartha, K.K. (1994). Self-fertile transgenic wheat plants regenerated from isolated scutellar tissues following bombardment with two distinct constructs. *Plant J.* **5**: 285–297.

Nikolov, D.B., Hu, S.H., Lin, J., Gasch, A., Hoffmann, A., Horikoshi, M., Chua, N.H., Roeder, R.G. and Burley, S.K. (1992). Crystal structure of TFII TATA-box binding protein. *Nature* **360**: 40–46.

Nishikawa, K. and Ishimaru, K. (1997). Flavonoids in root cultures of *Scutellaria baicalensis. J. Plant Physiol.* **151**: 633–636.

Nitsch, J.P. (1969). Experimental androgenesis in plants. *Phyto-morphology* **19**: 389–404.

Oksman-Caldentey, K.-M. and Arroo, R. (1999). Regulation of tropane alkaloid metabolism in plants and plant cell cultures. In: *Metabolic*

Engineering of Plant Secondary Metabolism. (Verpoorte, R. and Alfermann, A.W., eds.) Kluwer Acad. Pub., pp. 213–281.

Oksman-Caldentey, K.-M. and Hiltunen, R. (1996). Transgenic crops for improved pharmaceutical products. *Field Crops Res.* **45**: 57–69.

Oksman-Caldentey, K.-M., Kivelä, O. and Hiltunen, R. (1991). Spontaneous shoot organogenesis and plant regeneration from hairy root cultures of *Hyoscyamus muticus*. *Plant Science* **78**: 129–136.

O'Neill, C., Horvath, G.V., Horvat, E., Dix, P.Y. and Medgyesy, P. (1993). Chloroplast transformation in plants: Polyethylene glycol (PEG) treatment of protoplasts is an alternative to biolistic delivery system. *Plant J.* **3**: 729–738.

Owen, M., Gandecha, A., Cockburn, B. and Whitelam, G. (1992). Synthesis of a functional anti-phytochrome single-chain F_v protein in transgenic tobacco. *Bio/Technology* **10**: 790–794.

Parmenter, D.L., Boothe, J.G., Van Rooijen, G.J.H., Yeung, E.C. and Moloney, M.M. (1995). Production of biologically active hirudin in plant seeds using oleosin partitioning. *Plant Mol. Biol.* **29**: 1167–1180.

Parr, A.J., Peerless, A.C.J., Hamill, J.D., Walton, N.J., Robins, R.J. and Rhodes, M.J.C. (1988). Alkaloid production by transformed root cultures of *Catharanthus roseus*. *Plant Cell Rep.* **7**: 309–312.

Payne, J., Hamill, J.D., Robins, R.J. and Rhodes, M.J.C. (1987). Production of hyoscyamine by hairy root cultures of *Datura stramonium*. *Planta Medica* **53**: 474–478.

Perl, A., Kless, H., Blumenthal, A., Galili G. and Galun E. (1992). Improvement of plant regeneration and *gus* expression in scutellar wheat calli by optimization culture conditions and DNA-microprojectile delivery procedures. *Mol. Gen. Genet.* **235**: 279–284.

Plückthun, A. (1991). Antibody engineering. *Curr. Opin. Biotech.* **2**: 238–246.

Porceddu, A., Falorni, A., Ferradini, N., Cosentino, A., Calcinaro, F., Faleri, C., Cresti, M., Lorenzetti, F., Brunetti, P. and Pezzotti, M. (1999). Trangenic plants expressing human glutamic acid decarboxylase (GAD65), a major autoantigen in insulin-dependent diabetes mellitus. *Mol. Breeding* **5**: 553–560.

Porta, C. and Lomonossoff, G.P. (1996). Use of viral replicons for the expression of genes in plants. *Mol. Biotechnol.* **5**: 209–221.

Porta, C. and Lomonossoff, G.P. (1998). Scope for using plant viruses to present epitopes from animal pathogens. *Rev. Med. Virol.* **8**: 25–41.

Porta, C., Spall, V.E., Loveland, J., Johnson, J.E., Barker, P.J. and Lomonossoff, G.P. (1994). Development of cowpea mosaic virus as a high-yielding system for the presentation of foreign peptides. *Virology* **202**: 949–955.

Porta, C., Spall, V.E., Lin, T., Johnson, J.E. and Lomonossoff, G.P. (1996). The development of cowpea mosaic virus as a potential source of novel vaccines. *Intervirology* **39**: 79–84.

Porter, R.R. (1959). The hydrolysis of rabbit γ-globulin and antibodies with chrystallin papain. *Biochem. J.* **73**: 119–126.

Potrykus, I. and Spangenberg, G. (1995). *Gene Transfer to Plants.* Springer-Verlag, Berlin, 361 pp.

Pyke, K.A. (1999). Plastid division and development. *Plant Cell* **11**: 549–556.

Ramachandran, S., Hiratsuka, M.K. and Chua, N.H. (1994). Transcription factors in plant growth and development. *Curr. Opin. Genet. Dev.* **4**: 642–646.

Raven, P.H. and Johnson G.B. (1996). *Biology* (4th edition). Wm. C. Brown Pub. Dubuque, IA, USA.

Raybould, A.F. (1999). Transgenes and agriculture — Going with the flow. *Trends Plant Science* **4**: 246–248.

Reynolds, P.H.S. (ed.) (1999). *Inducible Gene Expression in Plants*. BCABI Publishing, Wallingford (UK), 247 pp.

Richter, L. and Kipp, P.B. (1999). Transgenic plants is edible vaccines. In: *Plant Biotechnology —New Products and Applications* (Hammon, J., McGravey, P. and Yusibov, V., eds.) Springer-Verlag, Heidelberg, pp. 159–176.

Richter, L., Mason, H.S. and Arntzen, C.J. (1996). Transgenic plants created for oral immunization against diarrheal diseases. *J. Travel Med.* **3**: 52–56.

Riker, A.J., Benfield, W.M., Wright, W.H., Keitt, G.W. and Sagen, H.E. (1930). Studies on infectious hairy roots on nursery apple tree. *J. Agr. Res.* **41**: 507–540.

Roitt, I., Brostoff, J. and Male, D. (1998). *Immunology* (5th edition). Mosby International, London.

Roth, D. and Craig, N.L. (1998). VDJ recombination: A transposase goes to work. *Cell* **94**: 411–414.

Saito, K., Noji, M., Ohmori, S., Imai, Y. and Murakoshi, I. (1991). Integration and expression of a rabbit liver cytochrome P-450 gene in transgenic *Nicotiana tabacum*. *Proc. Natl. Acad. Sci. USA* **88**: 7041–7045.

Saito, K., Kaneko, H., Tyamazaki, M., Yoshida, M. and Murakoshi, I. (1990). Stable transfer and expression of chimeric genes in licorice (*Glycyrrhiza uralensis)* using an Ri plasmid binary vector. *Plant Cell Rep.* **8**: 718–721.

Saito, K., Yamazaki, M., Shimomura, K., Yoshimatsu, K. and Murakoshi, I. (1990b). Genetic transformation of foxglove (*Digitalis purpurea)* by chimeric foreign genes and production of cardioactive glycosides. *Plant Cell Rep.* **9**: 121–124.

Salmon, V., Legrand, D., Slomianny, M.-C., Yazidi, I.E., Spik, G., Gruber, V., Bournat, P., Olagnier, B., Mison, D., Theisen, M. and

Merot, B. (1998). Production of human lactoferrin in transgenic tobacco plants. *Protein Exp. Purific.* **13**: 127–135.

Sanford, J.C., Klein, T.M., Wolf, E.D. and Allen, N. (1987). Delivery of substances into cells and tissues using a particle bombardment process. *J. Particle Sci. Tech.* **5**: 27–37.

Sardana, R.K., Ganz, P.R., Dudani, A., Tackaberry, E.S., Cheng, X. and Altosaar, I. (1998). Synthesis of recombinant human cytokine GM-CSF in the seeds of transgenic tobacco plants. In: *Methods in Biotechnology Vol. 3: Recombinant Proteins from Plants*. (Cunningham, C. and Porter, A.J.R., eds.) Humana Press, Totowa, NY, pp. 77–86.

Sasaki, K., Udagawa, A., Ishimara, H., Hayashi, T., Alfermann, A.W., Nakamishi, F. and Shimomura, K. (1998). High forskolin production in hairy roots of *Coleus forskohlii*. *Plant Cell Rep.* **17**: 457–459.

Sasson, A. (1998). Plant biotechnology-derived products: Market value estimates and public acceptance (based on lecture at IX Int. Cong. Plant Tissue & Cell Culture, Jerusalem). Kluwer Acad. Pub. Dortrecht, 160 pp.

Sauerwein, M., Yamazaki, T. and Shimomura, K. (1991). Hernandulcin in hairy root cultures of *Lippia dulcis*. *Plant Cell Rep.* **9**: 575–581.

Sauerwein, M. and Wink, M. (1993). On the role of opines in plants transformed with *Agrobacterium rhizogenes:* Tropane alkaloid metabolism, insect-toxicity and allelopathic properties. *J. Plant Physiol.* **142**: 446–451.

Schouten, A., Roosien, J., van Engelen, F.A., de Jong, G.A.M. (Ineke)., Borst-Vrenssen (Tanja) A.W.M., Zilverentant, J.F., Bosch, D., Stiekema, W.J., Gommers. F.J., Schots, A. and Bakker, J. (1996). The C-terminal KDEL sequence increaes the expression level of a single-chain antibody designed to be targeted to both the cytosol and the secretory pathway in transgenic tobacco. *Plant Mol. Biol.* **30**: 781–793.

Schuster, G., Lisitsky, I. and Klaff, P. (1999). Polyadenylation and degradation of mRNA in the chloroplast. *Plant Physiol.* **120**: 937–944.

Seding, J.M. (1998). Can ends justify the means? Telomeres and mechanisms of replicable senescence and immortalization in mammal cells. *Proc. Natl. Acad. Sci. USA* **95**: 9078–9081.

Sehnke, P.C. and Ferl, R.J. (1999). Processing of preproricin in transgenic tobacco. *Protein Exp. Purific.* **15**: 188–195.

Sehnke, P.C., Pedrosa, L., Paul, A.-L., Frankel, A.E. and Ferl, R.J. (1994). Expression of active, processed ricin in transgenic tobacco. *J. Biol. Chem.* **269**: 22473–22476.

Sevón, N., Hiltunen, R. and Oksman-Caldentey, K.M. (1998). Somaclonal variation in transformed roots and protoplast-derived hairy root clones of *Hyoscyamus muticus. Planta Medica* **64**: 37–41.

Shanks, J.V. and Morgan, J. (1999). Plant "hairy root" culture. *Curr. Opin. Biotech.* **10**: 151–155.

Shimomura, K., Sauerwein, M. and Ishmuru, K. (1991b). Tropane alkaloids in the adventitious and hairy root cultures of solanaceous plants. *Phytochem.* **30**: 2275–2278.

Shimomura, K., Sudo, H., Saga, H., Kamada, H. (1991a). Shikonin production and secretion by hairy root cultures of *Lithospermum erythrorhizon. Plant Cell Rep.* **10**: 282–285.

Sidkar, S.R., Serino, G., Chaudhuri, S. and Maliga, P. (1998). Plastid transformation in *Arabidopsis thaliana. Plant Cell Rep.* **18**: 20–24.

Sijmons, P.C., Dekker, B.M.M., Schrammeijer, B., Verwoerd, T.C., Van den Elzen, P.J.M. and Hoekema, A. (1990). Production of correctly processed human serum albumin in transgenic plants. *Bio/ Technology* **8**: 217–221.

Simpson, G.G. and Filipowicz, W. (1996). Splicing of precursors to mRNA in higher plants: Mechanisms regulation and sub-nuclear organization of the spliceosomal machinery. *Plant Mol. Biol.* **32**: 1–41.

Singer, M. and Berg, P. (1991). *Genes and Genomes. University Sci. Books*, Mill Valley, CA, 929 pp.

Smith, E.F. and Townsend, C.O. (1907). A plant tumour of bacterial origin. *Science* **25**: 671–673.

Sommer, S., Köhle, A., Yazaki, K., Shimomera, K., Bechthold, A. and Heide, L. (1999). Genetic engineering of shikonin biosynthesis hairy root cultures of *Lithospermum erythrorhizon* transformed with bacterial *ubi*C gene. *Plant Mol. Biol.* **39**: 683–693.

Spall, V.E., Porta, C., Taylor, K.M., Lin, T., Johnson, J.E. and Lomonossoff, G.P. (1997). Antigen expression on the surface of a plant virus for vaccine production. In: *Engineering Crops for Industrial End Use.* (Shewry, P.R., Napier, J.A. and Davis, P., eds.) Portland, London, pp. 35–46.

Srivastava, V., Anderson, O.D. and Ow, D.W. (1999). Single copy transgenic wheat regenerated through the resolution of complex integration patterns. *Proc. Natl. Acad. Sci. USA* **96**: 11117–11121.

Stachel, S.E. and Nester, E.W. (1986). The genetic and trancriptional organization in the *vir* region of the A6 Ti plasmid of *Agrobacterium tumefaciens. EMBO J.* **5**: 1445–1454.

Staczek, J., Bendahmane, M., Gilleland, L.B., Beachy, R.N. and Gilleland, H.E. Jr. (2000). Immunization with a chimeric tobacco mosaic virus containing an epitope of outer membrane protein F of *Pseudomonas aeruginosa* provides protection against challenge with *P. aeruginosa. Vaccine* **18**: 2266–2274.

Staub, J.M., Gracia, B., Graves, J. Hajdukiewicz, P.T.J., Hunter, P., Nehra, N., Parakar, V., Schlitter, M., Caroll, J.A., Spatola, L., Ward, D., Ye, G. and Russel, D.A. (2000). High-yield production of a human therapeutic protein in tobacco chloroplasts. *Nature Biotech.* **19**: 333–338.

Staub, J.M. and Maliga, P. (1992). Long regions of homologous DNA are incorporated into the tobacco plastid genome by transformation. *Plant Cell* **4**: 39–45.

Staub, J.M. and Maliga, P. (1993). Accumulation of D1 polypeptide in tobacco plastids is regulated via the untranslated region of the *psb*A mRNA. *EMBO J.* **12**: 601–606.

Subroto, A.M. and Doran, P.M. (1994). Production of steroidal alkaloids by hairy roots of *Solanum aviculare* and the effect of gibberellic acid. *Plant Cell Tissue Organ Culture* **38**: 93–102.

Sugiura, M. (1992). The chloroplast genome. *Plant Mol. Biol.* **19**: 149–168.

Sugiyama, Y., Hamamoto, Hi., Takemoto, S., Watanabe, Y. and Okada, Y. (1995). Systemic production of foreign peptides on the particle surface of tobacco mosaic virus. *FEBS Lett.* **359**: 247–250.

Sun, F.-L. and Elgin, S.C.R. (1999). Putting boundaries on silence. *Cell* **99**: 459–462.

Svab, Z., Hajdukiewicz, P. and Maliga, P. (1990). Stable transformation of plastids in higher plants. *Proc. Natl. Acad. Sci. USA* **87**: 8526–8530.

Svab, Z. and Maliga, P. (1993). High-frequency plastid trasnformation in tobacco by selection for a chimeric *aad*A gene. *Proc. Natl. Acad. Sci. USA* **90**: 913–917.

Tackaberry, E.S., Dudani, A.K., Prior, F., Tocchi, M., Sardana, R., Altosaar, I. and Ganz, P.R. (1999). Development of biopharmaceuticals in plant expression systems: cloning, expression and immunological reactivity of human cytomegalovirus glycoprotein B (UL55) in seeds of transgenic tobacco. *Vaccine* **17**: 3020–3029.

Tacket, C.O., Mason, H.S., Losonsky, G., Clements, J.D., Levine, M.M. and Arntzen, C.J. (1998). Immunogenicity in humans of a recombinant bacterial antigen delivered in a transgenic potato. *Nature Medicine* **4**: 607–609.

Takebe, K. and Hagiwara, K. (1998). Expression of human α-lactalbumin in transgenic tobacco. *J. Biochem.* **123**: 440–444.

Tavladoraki, P., Benvenuto, E., Trinca, S., De Martinis, D., Cattaneo, A. and Galeffi, P. (1993). Transgenic plants expressing a functional single-chain F_v antibody are specifically protected from virus attack. *Nature* **366**: 469–472.

Taylor, K.M., Porta, C., Lin, T., Johnson, J.E., Barker, P.J. and Lomonossoff, G.P. (1999). Position-dependent processing of peptides presented on the surface of cowpea mosaic virus. *Biol. Chem.* **380**: 387–392.

Tepfer, D. (1984). Transformation of several species of higher plants by *Agrobacterium rhizogenes*: Sexual transmission of the transformed genotype and phenotype. *Cell* **37**: 959–967.

Tepfer, D. (1995a). *Agrobacterium rhizogenes,* a natural transformation system. In: *Gene Transfer to Plants* (Potrykus, I and Spanenberg, G., eds.) Springer-Verlag, Berlin, pp. 41–44.

Tepfer, D. (1995b). *Agrobacterium-rhizogenes*-mediated transformation: Transformed roots to transformed plants. In: *Gene Transfer to Plants* (Potrykus, I. and Spanenberg, G., eds.) Springer-Verlag, Berlin pp. 45–52.

Thanavala, Y., Yang, Y.F., Lyons, P., Mason, H.S. and Arntzen, C. (1995). Immunogenicity of transgenic plant-derived hepatitis B surface antigen. *Proc. Natl. Acad. Sci. USA* **92**: 3358–3361.

Tingay, S., McElroy, D., Kalla, R., Fieg, S., Wang, M., Thornton, S. and Brettell, R. (1997). *Agrobacterium tumefaciens*-mediated barley transformation. *Plant J.* **11**: 1369–1376.

Trypsteen, M., van Lijsebettens, M., van Severen, R. and van Montagu, M. (1991). *Agrobacterium rhizogenes*-mediated transformation of *Echinacea purpurea. Plant Cell Rep.* **10**: 85–89.

Usha, R., Rohll, J.B., Spall, V.E., Shanks, M., Maule, A.J., Johnson, J.E. and Lomonossoff, G.P. (1993). Expression of an animal virus antigenic site on the surface of a plant virus particle. *Virology* **197**: 366–374.

Vandekerckhove, J., Van Damme, J., Van Lijsebettens, M., Botterman, J., De Block, M., Vandewiele, M., De Clercq. A., Leemans, J., Van Montagu, M. and Krebbers, E. (1989). Enkephalins produced in transgenic plants using modified 2S seed storage proteins. *Bio/Technology* **7**: 929–932.

Van Engelen, F.A., Schouten, A.L., Molthoff, J.W., Roosien, J., Salinas, J., Dirkse, W.G., Schots, A., Bakker, J., Gommers, F.J., Jongsma, M.A., Dirk, B. and Stiekemam W.J. (1994). Coordinate expression of antibody subunit genes yields high levels of functional antibodies in roots of transgenic tobacco. *Plant Mol. Biol.* **26**: 1701–1710.

Van Vloten-Doting, L., Boo, J.F. and Cornelissen, B. (1985). Plant-virus-based vectors for gene transfer will be of limited use because of high error frequency during viral RNA synthesis. *Plant Mol. Biol.* **4**: 323–326.

Vaquero, C., Sack, M., Changler, J., Drossard, J., Schuster, F., Monecke, M., Schillberg, S. and Fischer, R. (1999). Transient expression of a tumor-specific single-chain fragment and chimeric antibody in tobacco leaves. *Proc. Natl. Acad. Sci. USA* **96**: 11128–11133.

Vaucheret, H., Beclin, C., Elmayan, T., Feuerbach, F., Godon, C., Morel, J.-B., Mourrain, P., Palauqui, J.-C. and Vernhettes, S. (1998). Transgene-induced gene silencing in plants. *Plant J.* **16**: 651–659.

Velander, W.H., Johnson, J.L., Page, R.I., Russel, C.G., Subramanian, A., Wilkins, T.D., Gwazduaskas, F.C., Pittius, C. and Drohan, W.N. (1992). High level expression in the milk of transgenic swine using the cDNA encoding human protein C. *Proc. Natl. Acad. Sci. USA* **89**: 12003–12007.

Vitale, A. and Denecke, J. (1999). The endoplasmic reticulum — Gateway of the secretory pathway. *Plant Cell* **11**: 615–627.

Voss, A., Niersbach, M., Hain, R., Hirsch, H.-J., Liao, Y.-C., Kreuzaler, F. and Fischer, R. (1995). Reduced virus infectivity in *N. tabacum* secreting a TMV-specific full-size antibody. *Mol. Breeding* **1**: 39–50.

Wakasugi, T, Nagai, T., Kapoor, M., Sugita, M., Ito, M., Ito, S., Tsuduki, J., Nakashima, K., Tsudzuki, T., Suzuki, Y., Hamada, A., Ohta, T., Inamura, A., Yoshinaga, K. and Sugiura, M. (1997). Complete nucleotide sequence of the chloroplast genome from the green alga *Chlorella vulgaris:* The existence of genes possibly involved in chloroplast division. *Proc. Natl. Acad. Sci. USA* **94**: 5978–5972.

Wakasugi, T., Sugita, M., Tzudzuki, T. and Sugiura, M. (1998). Updated gene map of tobacco chloroplast DNA. *Plant Mol. Biol. Rep.* **16**: 231–241.

Wassenegger, M. and Pelisser, T. (1999). Signalling in gene silencing. *Trends Plant Science* **4**: 207–209.

Waterhouse, P.M., Smith, NA. and Wang, M.-B. (1999). Virus resistance and gene silencing: Killing the messenger. *Trends Plant Sci.* **4**: 452–458.

Watson, J.D. and Crick, F.H.C. (1953). A structure for deoxyribose nucleic acid. *Nature* **171**: 737.

Weeks, J.T. anderson, O.D. and Blechl, A.E. (1993). Rapid production of multiple independent lines of fertile transgenic wheat (*Triticum aestivum*). *Plant Physiol.* **102**: 1077–1084.

Wigdorovitz, A., Carrillo, C., Dus Santos, M.J., Trono, K., Peralta, A., Gomez, M.C. Rios, R.D., Franzone, P.M., Sadir, A.M., Escribano, J.M. and Borca, M.V. (1999). Induction of a protective antibody responsive to foot and mouth disease virus in mice following oral or parental immunization with alfalfa transgenic plants expressing the viral structural protein VP1. *Virology* **255**: 347–363.

Winter, G. and Milstein, C. (1991). Man-made antibodies. *Nature* **349**: 293–299.

Wongsamuth, R. and Doran, P.M. (1997). Production of monoclonal antibodies by tobacco hairy roots. *Biotech. Bioengin.* **54**: 401–415.

Wu, H., McCormac, A.C., Elliott, M.C. and Chen, D.-F. (1998). *Agrobacterium*-mediated stable transformation of cell suspension cultures of barley (*Hordeum vulgare*). *Plant Tissue Org. Culture* **54**: 161–171.

Wysokinska, H. and Rozga, M. (1998). Establishment of transformed root cultures of *Paulownia tomentosa* for verbascoside production. *J. Plant Physiol.* **152**: 78–83.

Ye, X., Al-Babili, S., Klotl, A., Zhang, J., Lucca, P., Beyer, P. and Potrykus, I. (2000). Engineering the provitamin A (β-carotene) biosynthetic pathway into (carotenoid-free) rice endosperm. *Science* **287**: 303–305.

Yoon, J.W., Yoon, C.S., Lim, H.W., Huang, Q.Q., Kang, Y., Pyun, K.H. Hirasawa, K., Sherwin, R.S. and Jun, H.S. (1999). Control of autoimmune diabetes in NOD mice by GAD expression or suppression in beta cells. *Science* **284**: 1183–1187.

Yoshikawa, T. and Furuya, T. (1987). Saponin production by cultures of *Panax ginseng* transformed with *Agrobacterium rhizogenes. Plant Cell Rep.* **6**: 449–453.

Yukimune, Y., Hara, Y. and Yamada, Y. (1994). Tropane alkaloid production in root cultures of *Duboisia myoporoides* obtained by repeated selection. *Biosci. Biotech. Biochem.* **58**: 1443–1446.

Yun, D.-J., Hashimoto, T. and Yamada, Y. (1992). Metabolic engineering of medicinal plants: Transgenic *Atropa belladonna* with an improved alkaloid composition. *Proc. Natl. Acad. Sci. USA* **89**: 11799–11803.

Yusibov, V., Modelska, A., Steplewski, K., Agadjanyan, M., Weiner, D., Hooper, D.C. and Koprowski, H. (1997). Antigens produced in plants by infectoin with chimeric plant viruses immunize against rabies virus and HIV-1. *Proc. Natl. Acad. Sci. USA* **94**: 5784–5788.

Yusibov, V., Shirprasad, S., Turpen, T.H., Dawson, W. and Korpowski, H. (1999). Plant viral vectors based on tobamoviruses. In: *Plant Biotechnology — New Products and Applications* (Hammon, J., McGravey, P. and Yusibov, V., eds.) Springer-Verlag, Heidelberg, pp. 81–94.

Zambryski, P. (1997). Genetic manipulation of plants. In: *Exploring Genetic Mechanisms* (Singer, M. and Berg, P., eds.) University Sci. Books, Sausalito, CA, pp. 597–655.

Zeitlin, L., Olmsted, S.S., Moench, T.R., Co, M.S., Bartinell, B.J., Paradkar, V.M., Russel, D.R., Queen, C., Cone, R.A. and Whaley, K.J. (1998). A humanized monoclonal antibody produced in transgenic plants for immunoprotection of the vagina against genital herpes. *Nature Biotech.* **16**: 1361–1364.

Zhong, G.-Y., Peterson, D., Delaney, D.E., Bailey, M., Witcher, D.R., Register, J.C., Bond, D., Li, C.-P., Marshall, L., Kulisek, E., Ritland, D., Meyer, T., Hood, E.E. and Howard, J.A. (1999). Commercial production of aprotinin in transgenic maize seeds. *Mol. Breeding* **5**: 345–356.

Zhou D.-X. (1999). Regulatory mechanism of plant gene transcription by GT-elements and GT factors. *Trends Plant Science* **4**: 210–214.

Zhu, Z., Hughes, K.W., Huang, L., Sun, B., Liu, C., Li, Y., Hour, Y. and Li, X. (1994). Expression of human α-interferon cDNA in transgenic rice plants. *Plant Cell Tissue Organ Culture* **36**: 197–204.

Zimmerman, S., Schillberg, S., Liao, Y.-C. and Fisher, R. (1998). Intracellular expression of TMV-specific single-chain F_v fragments leads to improved virus resistance in *Nicotiana tabacum*. *Mol. Breeding* **4**: 369–379.

Index